Global Environmental Studies

Editor-in-Chief

Ken-ichi Abe, Research Institute for Humanity and Nature, Kyoto, Japan

Series Editors

Daniel Niles, Research Institute for Humanity and Nature, Kyoto, Japan

Hein Mallee, Kyoto Prefectural University, Kyoto, Japan

This series introduces the research undertaken at, or in association with, the Research Institute for Humanity and Nature. Located in Kyoto, Japan, RIHN is a national institute conducting fixed-term, multidisciplinary and international research projects on pressing areas of environmental concern. RIHN seeks to transcend the common divisions between the humanities and the social and natural sciences, and to develop synthetic and transformative description of humanity in the midst of a dynamic, changeable nature. The works published here will reflect the full breadth of RIHN scholarship in this transdisciplinary field of global environmental studies.

Kosuke Mizuno • Osamu Kozan • Haris Gunawan
Editors

Vulnerability
and Transformation
of Indonesian Peatlands

 Springer

Editors
Kosuke Mizuno
School of Environmental Science
University of Indonesia
Jakarta, Indonesia

Center for Southeast Asian Studies
Kyoto University
Kyoto, Japan

Haris Gunawan
University of Riau
Pekanbaru
Riau, Indonesia

Osamu Kozan
Center for Southeast Asian Studies
Kyoto University
Kyoto, Japan

This work was supported by the Research Institute for Humanity and Nature

ISSN 2192-6336 ISSN 2192-6344 (electronic)
Global Environmental Studies
ISBN 978-981-99-0908-7 ISBN 978-981-99-0906-3 (eBook)
https://doi.org/10.1007/978-981-99-0906-3

This Springer imprint is published by the registered company Springer Nature Singapore Pte Ltd.
The registered company address is: 152 Beach Road, #21-01/04 Gateway East, Singapore 189721, Singapore

Preface

The degradation of tropical peat swamps in Southeast Asia has increasingly become problematic in the context of international environmental conservation. Large-scale drainage associated with plantation development of fast-growing trees and oil palm has led to a decrease in groundwater table levels and the drying of peat swamp forests.

This has resulted in an increase in CO_2 emissions by peat decomposition, and frequent fires that have significantly contributed to global warming. The attendant haze has caused incalculable damage to the local economy and impacted the health of not only the local people but also those of neighboring countries. The prevention of peatland degradation and aridification has become a global environmental, economic, and political concern. Over the last several years, there have been several efforts to prevent these fires and restore degraded peatlands.

Today, there are scores of studies on peatland degradation and restoration from various viewpoints such as carbon emission, fire, haze, water management, revegetation, and restoration.

This book attempts an integrated approach toward understanding, reconstructing, and reshaping the issue of peatland degradation and regeneration by employing the ideas of vulnerability, resilience, adaptability, and transformation of the Indonesian peatland ecosystem and its relationship with society.

Although peat swamp forests have been resilient, they remain vulnerable to development that involves drainage, such as in large-scale plantations. There might be many lessons at the peat swamp forest that had lasted for thousands of years to keep the resilience and adaptability of peat ecosystem. To be sure, multiple generations of communities that have lived in the peat ecosystem have learned such precious lessons that contributed to the peatlands' adaptability. Ecological, biological, geological, and climatic conditions have supported its resilience as well. Also, there are certainly reasons for hope in the resilience of the new system in the current peatland crisis, such as people's efforts based on past experience to control fires, attempts to develop innovative paludicultural practices, and efforts to tap into the local knowledge of peat swamp forests to restore degraded peatlands and maintain ecological sustainability. Our studies on sociological, ecological, biological, geo-logical, and climatic conditions for peat swamp forests will support those initiatives.

Hence, in this book, we present our integrated method to demonstrate the vulnerability, resilience, adaptability, and transformation of peatlands to shed light on what truly constitutes "resilience" of the peat swamp forest, and in the coming third volume of our project titled "Toward the Regeneration of Tropical Peatland Societies: Building International Research Network on Paludiculture and Sustainable Peatland Management." We dig deeper into the local knowledge and develop our studies on institutions, governance, and ecological conditions that support the resilience of the peat swamp forest to discuss the idea of transformation in today's degraded peatlands.

This book examines a host of concerns relating to peatland degradation and rehabilitation through the integrated lens of vulnerability, resilience, adaptability, and transformation of the Indonesian peatland ecosystem. Here we focus on issues such as land titles, the bird community, Ground-Dwelling Mammals, biomass and carbon restoration of the peat swamp ecosystem, and CO_2 emission from drained peatland as part of vulnerability; termite and timber processing/marketing as part of resilience and adaptability; and the resetting, revegetation, and revitalization of livelihood as well as ISPO certification process as part of transformation.

Many of us started our fieldwork in Sumatra in 2008 in the project titled "In Search of Sustainable Humanosphere in Asia and Africa" (2007–12) for the Kyoto University Global Center of Excellence program.

At the start of our fieldwork, peatland degradation was in dire straits. As peatland issues are related to several other sociocultural and economic aspects, we employed a multidisciplinary and interdisciplinary approach. These fieldworks have yielded publications deploying the idea of a sustainable humanosphere, for example, *Catastrophe and Regeneration in Indonesia's Peatlands: Ecology, Economy and Society* (Mizuno et al. ed. 2016 published by NUS Press and Kyoto University Press).

We have developed our discussion to include the period of pre-degradation and several efforts at peatland restoration for a better understanding and analysis of the long-term peatland dynamics. We also employ the ideas of vulnerability, resilience, adaptability, and transformation based on expanded studies on peatlands, and observations of and participation in multiple efforts to prevent fires and restore the degraded peatland by researchers, the government, nongovernment organizations (NGOs), private companies, and last but not least, the local people.

The restoration of degraded peatlands to help mitigate global warming discussed in this book directly relates to SDG 15 and SDG 13. SDG 15 is a call to *protect, restore, and promote sustainable use of terrestrial ecosystems, sustainably manage forests, combat desertification, and halt and reverse land degradation and halt biodiversity loss.* Peatlands originally formed the peat swamp forest, hence SDG 15.2, which *promotes the sustainable management of all types of forests, halts deforestation, restores degraded forests, and increases afforestation and reforestation globally*, is closely related. When peatlands were peat swamp forests containing huge amounts of carbon and water and constituted a habitat for rare species of flora and fauna, it was all well and good. But when human exploitation, accompanied by drainage, started, degradation set in, especially after the large-scale plantation of oil palm and timber, and brought about the intrusion of immigrants into the peat swamp forest area—all of which accelerated the degradation of peat swamp forests. Concerted efforts to halt peatland degradation over the last decades have raised questions on the sociological, economic, and ecological factors that contributed to peatland degradation, and how all stakeholders can promote peatland restoration and transformation.

SDG 13 is a call to *take urgent action to combat climate change and its impacts.* Drainage in peat swamps that contain huge amounts of carbon has resulted in an increase in CO_2 emissions through peat decomposition, as well as frequent fires that have significantly contributed to global warming. The people's initiative of rewetting can help curb peatland decomposition and prevent peatland fires, and significantly mitigate global warming. Peatland decomposition and fires emit massive amounts of carbon, which means that peatland restoration to prevent peatland fires would have a substantial impact on reducing the amount of carbon emissions.

We hope this book will contribute to a more nuanced appreciation of peatland issues and help further efforts to restore the peatlands toward improving the health and environmental conditions in the region for the welfare of local communities.

We are deeply grateful to Springer and the Research Institute for Humanity and Nature, which handled the publication of this book. We also recognize the Japan Society for the Promotion of Science (JSPS) for its support through the JSPS Grants-in-Aid for Scientific Research, Project No. 19H04350 ("Land ownership and peatland restoration in Indonesia") and RIHN for support of the research project "Toward the Regeneration of Tropical Peatland Societies: Building International Research Network on Paludiculture and Sustainable Peatland Management", 2017–2021.

We are also indebted to all the scholars, universities, governments, NGOs, companies, research institutes, and local people (who are too many to mention) that have been part of this prodigious exploration.

Jakarta, Indonesia/Kyoto, Japan   Kosuke Mizuno
Kyoto, Japan   Osamu Kozan
Riau, Indonesia   Haris Gunawan

Contents

Contributors

Muhammad Gevin Ardi Center for Disaster Studies, University of Riau, Pekanbaru, Riau, Indonesia

Aulia Aruan Alastair Fraser Forestry Foundation, Jakarta, Indonesia

Yudi Chadirin Department of Civil and Environmental Engineering, IPB University, Bogor, Indonesia

Chairul Chemical Engineering Department, Center for Disaster Studies, University of Riau, Pekanbaru, Riau, Indonesia

Motoko S. Fujita Center for Southeast Asian Studies, Kyoto University, Kyoto, Japan

Haris Gunawan University of Riau, Pekanbaru, Riau, Indonesia

Dendy Sukma Haryadi Bogor Agricultural University, Bogor, West Java, Indonesia

Masayuki Itoh School of Human Science and Environment, University of Hyogo, Himeji, Hyogo, Japan

Osamu Kozan Center for Southeast Asian Studies, Kyoto University, Kyoto, Japan

Kazuya Masuda Faculty of Agriculture and Marine Science, Kochi University, Nankoku-shi, Kochi, Japan

Kosuke Mizuno School of Environmental Science, University of Indonesia, Jakarta, Indonesia
Center for Southeast Asian Studies, Kyoto University, Kyoto, Japan

Ahmad Muhammad University of Riau, Pekanbaru, Riau, Indonesia

Toshihide Nagano Takasaki University of Health and Welfare, Takasaki, Gunma, Japan

Daisuke Naito Graduate School of Agriculture, Kyoto University, Kyoto, Japan

Besri Nasrul Soil Science Department, Center for Disaster Studies, University of Riau, Pekanbaru, Riau, Indonesia

Kok-Boon Neoh Department of Entomology, National Chung Hsing University, Taichung, Taiwan

Dian Novarina Asia Pacific Resources International Limited, Jakarta Pusat, Indonesia

Kazutoshi Osawa Utsunomiya University, Utsunomiya, Tochigi, Japan

Adhy Prayitno Mechanical Engineering Department, Center for Disaster Studies, University of Riau, Pekanbaru, Riau, Indonesia

Akhbar Putra Center for Disaster Studies, University of Riau, Pekanbaru, Riau, Indonesia

Rinaldi Civil Engineering Department, Center for Disaster Studies, University of Riau, Pekanbaru, Riau, Indonesia

Hiromitsu Samejima Institute for Global Environmental Strategies, Hayama, Kanagawa, Japan

Satyanto Krido Saptomo Department of Civil and Environmental Engineering, IPB University, Bogor, Indonesia

Budi Indra Setiawan Department of Civil and Environmental Engineering, IPB University, Bogor, Indonesia

Satomi Shiodera Center for Southeast Asian Studies, Kyoto University, Kyoto, Japan
Department of Global Liberal Studies, Faculty of Global Liberal Studies, Nanzan University, Nagoya, Aichi, Japan

Susilo Sudarman PT Riau Andalan Pulp and Paper, Pelalawan, Riau, Indonesia

Sigit Sutikno Civil Engineering Department, Center for Disaster Studies, University of Riau, Pekanbaru, Riau, Indonesia

Haruka Suzuki Faculty of Humanities and Sciences, Kobe Gakuin University, Kobe, Japan

Almasdi Syahza Institute of Research and Community Services, University of Riau, Pekanbaru, Riau, Indonesia

Hesti Lestari Tata Research Center of Ecology and Ethnobiology, National Research and Innovation Agency, Bogor, Indonesia

Dede Hendry Tryanto The Indonesian Environment Fund, Jakarta, Indonesia

Bondan Widyatmoko National Research and Innovation Agency, Jakarta, Indonesia

Yesi Sociology Science Department, Center for Disaster Studies, University of Riau, Pekanbaru, Riau, Indonesia

Muhamad Yusa Civil Engineering Department, Center for Disaster Studies, University of Riau, Pekanbaru, Riau, Indonesia

Chapter 1
Introduction: The Vulnerability and Transformation of Indonesian Peatlands

Kosuke Mizuno, Motoko S. Fujita, Osamu Kozan, Masayuki Itoh, Satomi Shiodera, Daisuke Naito, Haruka Suzuki, and Haris Gunawan

Abstract Tropical peat swamp forests that started to experience fires in the 1960s in Sumatra and in the 1980s in Kalimantan are now seriously degraded in many places in Indonesia. But from a longer perspective, we can see the tropical peat swamp forests' resilience and adaptation. There have been several reports on how deep local knowledge has helped curb further degradation of these forests. Large-scale commercial logging had been undertaken in the Riau area in Sumatra in the 1860s; however, no serious peatland degradation and large-scale fires were reported until the end of the colonial era. Truly serious degradation started only in the middle of the 1980s when large-scale drainage systems were created to support the extensive cultivation of oil palm and timber plantations. Serious fires broke out in 1997, and in 2015 at the time of El Niño. Today only 18.4% of total peatlands in Indonesia are undisturbed natural peatland forests. We also find several initiatives to restore

K. Mizuno (✉)
School of Environmental Science, University of Indonesia, Jakarta, Indonesia

Center for Southeast Asian Studies, Kyoto University, Kyoto, Japan
e-mail: mizuno@cseas.kyoto-u.ac.jp

M. S. Fujita · O. Kozan
Center for Southeast Asian Studies, Kyoto University, Kyoto, Japan

M. Itoh
School of Human Science and Environment, University of Hyogo, Himeji, Hyogo, Japan

S. Shiodera
Center for Southeast Asian Studies, Kyoto University, Kyoto, Japan

Department of Global Liberal Studies, Faculty of Global Liberal Studies, Nanzan University, Nagoya, Aichi, Japan

D. Naito
Graduate School of Agriculture, Kyoto University, Kyoto, Japan

H. Suzuki
Faculty of Humanities and Sciences, Kobe Gakuin University, Kobe, Japan

H. Gunawan
University of Riau, Pekanbaru, Riau, Indonesia

1
K. Mizuno et al. (eds.), *Vulnerability and Transformation of Indonesian Peatlands*, Global Environmental Studies, https://doi.org/10.1007/978-981-99-0906-3_1

degraded peatlands. For one thing, the government has established the Peatland Restoration Agency (BRG) to coordinate the acceleration of peatland restoration and promote the welfare of local people in the peatland area. For another, international organizations, companies, nongovernment organizations, and local communities also have started their respective programs to rehabilitate the peatlands. Such developments can be interpreted as attempts at transformation, that is, the creation of a fundamentally new system in response to ecological, economic, or social (including political) conditions that render the existing system untenable. This volume intends to reshape the discussion on peatland degradation and regeneration with the ideas of vulnerability, resilience, adaptability and transformation, and deepen the understanding of degradation and regeneration.

Keywords Tropical peat swamp forest · Vulnerability · Resilience · Adaptability · Transformation

1.1 Introduction

Tropical peat accumulation was a primary feature of the Holocene, as well as of the last glacial period. Peat accumulation in most tropical coastal areas in Southeast Asia commenced around 4000–5000 BP (Page et al. 2006; Anderson and Muller 1975). The subcoastal and inland peatlands in Borneo date back to the late Pleistocene (ca. 29,000 BP) (Page et al. 2006; Anshari et al. 2001, 2004).

Tropical peatlands store a huge amount of carbon. Tropical peatlands may account for only 10–12% of the global peatland resource by area, but owing to their considerable thickness and high carbon content, they store between 50 Gt and 70 Gt (16–21%) of peat soil carbon, or about 2–3% of the carbon stored in all soils globally (Page et al. 2002, 2006).

Tropical peat swamp forests in their natural state make an important contribution to regional and global biodiversity (Andriesse 1988; Page and Rieley 1998) providing a vital, but undervalued, habitat for rare and threatened species, especially birds, mammals, reptiles and freshwater fish (Ismail 1999; Thornton et al. 2020). "Until recently, peat swamps were assumed to be hostile, acidic places where the biodiversity was low." "Officials and developers argue that there is no point conserving the swamps because there is 'nothing' there. These places don't have big, sexy animals, but in almost all cases, when they say a place is species-poor, they're wrong" (Dennis and Aldhous 2004, pp. 396, 398). Posa et al. (2011) documented 1524 plant species in peat swamp forests in Southeast Asia, of which 172 were endemic species, and 219 species of freshwater fish of which 80 species were endemic.

Peatlands have been exploited for a long time, starting with small-scale clearing and collection of nontimber products by local people. Fisheries, small-scale shifting cultivation and rubber cultivation provided principal sources of livelihood for local people (Furukawa 1992). Commercial logging has also taken place in Riau, Sumatra, Indonesia, since the middle of the nineteenth century but did not seriously disturb the peat swamp forest ecosystem (Mizuno et al. 2021). The exploitation of peatlands for the purpose of oil palm and timber since 1980s, however, accompanied by construction of large-scale drainage projects, has seriously degraded peatlands, resulting in burning and abandonment by settlers.

1.2 Vulnerability of Peatlands

Peatlands ecosystems are vulnerable, especially to the changes wrought by construction of large-scale drainage systems (Mizuno et al. 2016). This book, which is the result of over 12 years of peatlands research conducted by Japanese and Indonesian scholars from a range of natural and social science disciplines, describes the many dimensions of vulnerability, resilience, adaptability, and transformation of Indonesian peatlands social and ecological systems.

In principle, peatlands are not suitable for cultivation because once exploited, they show characteristics of acid sulfate soil, as well as subsidence, disappearance and degradation (Furukawa 1992, pp. 25–31). Nevertheless, Indonesian peatlands have been the object of numerous exploitation or development policies. The construction of drainage systems has caused significant CO_2 emissions (Jaenicke et al. 2008) arising not just from the decomposition of peat (Agus et al. 2010) but also from the fires that have become common in the newly-desiccated landscapes (Page and Hooijer 2016; Purnomo et al. 2017, 2019). As reported by Page et al. (2002), during the 1997 El Niño event, 0.73 million hectares of peatland were burned. Haze from the peatland fires spread to neighboring countries (Varkkey 2013), causing health problems among local populations (Uda et al. 2019; Marlier et al. 2019). Emissions from deforestation due to drainage and fires have been the single largest source of greenhouse gas (GHG) emissions from land use in Southeast Asia. Total carbon loss from converting peat swamp forests into oil palm production zones is 59.4 ± 10.2 Mg of CO_2 per hectare per year during the first 25 years after land-use cover change, of which 61.6% is emitted from peat. Of the total amount (1486 ± 183 Mg of CO_2 per hectare over 25 years), 25% is released immediately from land-clearing fire (Murdiyarso et al. 2010).

Almost every form of peatland development has to some extent involved drainage of the peatland itself and/or its surrounding area (ASEAN 2007). Traditionally, farmers developed relatively small and closed-ended canals for drainage. Over the years many more canals were constructed for logging activities, both in order to drain peatland soils and to facilitate transport of the harvested logs. More systematic drainage began in the mid-1990s, led by the Mega Rice Project (van Beukering et al. 2008; Wösten et al. 2008) and companies establishing acacia and oil palm plantations (Wösten et al. 2008). Evers et al. (2016) have criticized the proposition that drainage-based agriculture can be a component of sustainable tropical peatland development. They argued that even where current guidance and/or policies are implemented to monitor water table depths (WTDs) and limit development to only certain proportions of peat domes, such "sustainable" practices inevitably result in significant GHG emission, long-term subsidence of peat soils, significant flooding in Southeast Asian coastal peatlands, as well as significant losses of biodiversity. The race to catalogue biodiversity before it disappears is particularly intense in the peat swamps, which are vanishing at a frightening rate (Dennis and Aldhous 2004).

1.3   Resilience and Adaptation of Peat Swamp Forest

Peat swamp forests can instead be a showcase for social and ecological resilience and adaptation. Resilience here refers to the capacity of a system to absorb distur-bance and reorganize while undergoing change so as to still retain essentially the same function, structure and feedbacks, and therefore identity. We therefore under-stand resilience as the capacity to change in order to maintain a relatively consistent identity. Adaptability (adaptive capacity) here refers to the capacity of actors in a system to influence resilience (Walker et al. 2004; Folke et al. 2010).

Momose (Momose 2002; Momose and Shimamura 2002) has vividly portrayed sustainable use of mixed peat swamp forests along riverbanks of the Kampar River in Riau Province. People in such areas have traditionally lived in close relationships with mixed peat swamp forest. Their material interactions involve hunting, gathering of many food and utilitarian species, and agriculture, especially swidden cultivation of rice. Spiritually they rely, for example, on the healing powers of plants and herbs found deep in the forest. They possess a rich knowledge of the local landscape and of the networks of trade, marriage, and migration that have sustained local lifeways. Such networks ensured that village peoples successfully adapted to environmental and social changes through time, including population increase, and avoided overexploitation of mixed peat swamp forests. The Malay people of this study have pushed into the frontier of the tidal zone, engaging in fishing, intensive agriculture, and the use of mixed peat swamp forests of the area. They extended their traditional networks into cities via trade. These linkages can be thought as relieving population pressure on peatlands.

Before the introduction of large-scale commercial logging, timber plantations, and intensive oil palm plantations there were no reports of serious damage from fires. The traditional *Ongka* logging (a method of logging using wood rails and sleighs) observed by the local people did not have the technical capability to rapidly expand the logging area, so there was some balance maintained between logging and tree growth (Momose 2002). The peat swamp fires began in the 1960s in Sumatra and then in Kalimantan in the 1980s. During periods of drought, such as in El Niño years, large fires broke out. In Kalimantan, large fires flared for the first times in 1982, 1992, and 1997, even though there had been severe droughts in 1962, 1964, and 1973. On the other hand, in Sumatra, peatlands became more sensitive to autumnal drought after 1985, as there were no significant fires documented between 1960 and 1984 (Field et al. 2009).

In Riau, Sumatra, where our studies were mainly concentrated, numerous large-scale fires were recorded since the early 1980s (Kieft et al. 2016). The mega-fires of

1997–1998 were the biggest fire events in Riau (and Indonesia). The Riau fires were attributed to the large-scale exploitation of peatlands to meet demand for peatland commodities such as oil palm and timber for pulp and paper (Saputra 2019).

This history of exploitation or peatland use demonstrates that the peat swamp forest was resilient at least until the 1960s. For example, during 1860–1940 there were thriving *Panglong* (logging) businesses in the Riau peatland area. *Panglong* was the main business in Riau and Riau islands in Sumatra, run by Chinese businessmen based in Singapore. Chinese laborers were recruited mainly from Singapore. Charcoal-making businesses run by local people were included in *Panglong* (Pastor 1927). The amount of timber logged in 1927 was 450,000 m³— the largest haul in the Netherlands Indies timber industry (Jelles 1929). There were concerns about depletion of timber resources at the time, but no reports of serious peatland degradation or peatland fires (Sewandono 1937).

Page et al. (2009) argue that the livelihood activities of the Dayak people in Kalimantan and the Malay people in Sumatra have been ecologically compatible with maintaining peatland natural resource functions with only limited and localized damage. There is evidence that peatland ecosystems could accommodate small-scale agriculture, sago planting, nontimber product collection, and fisheries throughout the colonial era (Reep 1907; Mizuno et al. 2021). Since 1965, however, the peat swamps in those areas have been subjected to rapid land-use changes and degradation as a result of transmigration (both spontaneous and officially-sanctioned), large-scale logging (legal and illegal), the establishment of oil palm plantations, decreasing rotation of timber plantations to supply the pulp and paper industries, and fire. By 1988, more than 93% of the peat swamp forests of Sumatra had been degraded, leaving only a few areas in primary ecological states (Page et al. 2009).

Today, peatlands in Indonesia are seriously degraded. According to Uda et al. (2017), of the total 14,915,000 hectares of peatland, undisturbed natural peatland forest comprises only 18.4% of the total, while disturbed natural forest amounts to 24.7%. Meanwhile degraded peatlands, including abandoned land, account for 26.9% of the total, oil palm forests 16.2%, and acacia plantation forest 16.2%. Of the remaining 24.5%, 6.2% is comprised of dry agricultural land, 2.4% of wet rice field, and there are a number of other uses.

1.4 Transformation-Peatland Restoration

To be sure, there have been many peatland restoration efforts. Rewetting has been given priority; canal blocking (Applegate 2012; Ritzema et al. 2014) and hydrological restoration have been discussed (Jaenicke et al. 2010). Rewetting is the key for rehabilitation, and paludiculture (Uda et al. 2020)—the practice of crop and timber planting on wet soils—must be undertaken on rewetted peatland. In her report on the developments of paludiculture, Tata (2019) reported that it was difficult to maintain water levels required by paludiculture, and that lower water tables led to subsidence and increased carbon emission. Dohong et al. (2018) discussed the technical aspects of peatland restoration, focusing on hydrological and vegetational restoration.

The Peatland Restoration Agency (*Badan Restorasi Gambut*, BRG) was established by presidential decree in 2016 to coordinate the acceleration of the restoration of peatlands and promote the welfare of local people in the peatland area. The BRG has advocated the 3R program—Rewetting, Revegetation, Revitalization of Livelihood. The 3R program has been implemented at Desa Peduli Gambut (Peatland Care Village). Today a number of 3R trial areas have been established in order to popularize the program. Studies show that these trial areas, for example in the sago palm area in Kepulauan Meranti District, Riau Province, are still limited in scope, however (Ehara et al. 2018).

There are also several peatland restoration initiatives led by international organizations, the private sector, and nongovernment organizations. Companies that secured ecosystem restoration concessions have attempted gradual revegetation in deforested areas, assisted natural regeneration, rewetting, community-based fire prevention, and established an agroecological farmer school (Darusman et al. 2021). They have also introduced a new method of rewetting called "stock-based water management," which functions like an irrigation system for the paddy field. This system is designed to achieve a high level of water retention in the ecosystem (Kato et al. 2021).

Such developments can be interpreted as attempts at transformation, that is, creation of a fundamentally new system in response to ecological, economic, or social (including political) conditions that render the existing system untenable (Walker et al. 2004; Folke et al. 2010). From our perspective on peatland ecosystems and livelihoods, this new system should not degrade peatlands and must create a sustainable ecosystem. It should guard against peatland fires and revive local livelihoods.

Transformations can be seen to consist of three phases (Olsson et al. 2004; Folke et al. 2010): (1) being prepared for, or even preparing, the social-ecological systems for change; (2) navigating the transition by using crisis as a window of opportunity for changes; and (3) building resilience of the new social-ecological regime. Transformational change often involves shifts in perception and meaning, social network configurations, patterns of interaction among actors, including leadership and politics and power relations, and associated organizational and institutional arrangement (Folke et al. 2009, 2010; Huitema and Meijerink 2009; Smith and Stirling 2010). The new system, once established, should have resilience and adaptability as described.

The current 3R program may thus be viewed as a kind of preparation for the socio-ecological system to come, or as a way of navigating the transition by making use of a crisis as an opportunity for change. Both the Indonesian government and president have repeatedly emphasized the importance of taking care of the peatlands and the urgency of fire prevention in peatland landscapes. The BRG diffused the idea through its Desa Peduli Gambut (Peatland Care Village) and MPA (*Masyarakat Peduli Api*, Fire Care Community) program, which will be explained in volumes to follow.

The peat swamp forest may have been resilient in the past, but as mentioned it is still vulnerable today. Transformation to a newly resilient socio-ecological state might depend in part on the lessons of peat swamp forest life that lasted for thousands of years. At present, the peatlands are facing a crisis, but there are reasons for hope in the resilience and adaptability of the new system, such as the people's initiatives to curb fires or trials in developing a new paludiculture, or tapping into local knowledge about the peat swamp forest to keep the ecosystem sustainable.

While there are numerous studies on peatland degradation and restoration from various viewpoints such as carbon emission, fire, haze, water management, revegetation, and restoration, an integrated idea toward understanding, reconstructing, and reshaping the issue of peatland degradation and regeneration is still lacking. Our long-term research project, therefore, presents an integrated approach to transformation of peatlands, investigating what "resilience" of peat swamp forest means from the perspective of ecological science, social science, and cultural livelihoods. Our project is fundamentally concerned with local knowledge on the resilience of the peat swamp forest, and works together with Indonesian scholars and communities in order to articulate the many dimensions of transformation of degraded peatlands.

1.5   Structure of the Book

The present volume discusses vulnerability, resilience, adaptability, and transformation. Part one focuses on social and ecological vulnerability. Chapter 2 by Mizuno et al. explicate the dynamics of vulnerability of the peat swamp forest in the research village of Tanjung Leban, Bengkalis District, Riau Province. The former local population comprising Malays had established convivial interactions with the peatland ecosystem. However, especially after an acacia plantation was established in the area, peatland fires began to break out. As a result, people turned to logging the peat swamp forest, and land they cleared was subsequently distributed among local people. Along with the development of the settlement and the increase in the number of migrants, people have acquired land through clearing, inheritance, and purchase, and began to secure land rights through letters of certification issued by village authorities. Those lands without letters tend to be abandoned after experiencing fires, on the other hand, lands with letters to certificate land rights tend not to be abandoned, although those land also experienced (some time repeated) fires.

Chapter 3 by Fujita et al. argue that peat swamp forest vulnerability can be understood in part through study of its bird population. Contrary to earlier assessments, Fujita et al. demonstrate that peat swamp forest has a high species variety and overall number of birds. Acacia plantation forests tend to have low numbers of birds, while intermediate zones of rubber forest and village forest have middle-level frequency of bird appearance. These data show the vulnerability of the acacia forest, and some degree of adaptability in the rubber and village forests, hinting at the new regime of transformation from the viewpoint of biodiversity.

Chapter 4 by Samejima et al. take a very similar approach, but this time from the perspective of mammals. Their study shows that the frequency of mammal presence in the peat swamp forest and the acacia forest follows a similar trajectory: high in peat swamp forest and low in acacia forest. This study therefore also points to the resilience of the peat swamp forest and the near term vulnerability of the acacia forest.

Chapter 5 by Saptomo et al. show that CO_2 emission from bare peatlands is affected by temperature and soil humidity. Lower temperature and higher soil humidity lessen CO_2 emission. Soil humidity is not easily controlled, and rainfall has a clear influence. This study therefore demonstrates the vulnerability of bare peatlands.

While all the foregoing papers demonstrate the resilience of the peat swamp forest and vulnerability of peatlands, especially after degradation or the spread of the acacia forest, the next three papers emphasize the peat swamp forest's resilience and adaptability. Chapter 6 by Kok-Boon Neoh et al. discuss termites in degraded and abandoned peatlands. Even as local people wage war against termites, which are typically seen as pests, the insects play a potentially important role in peatland restoration. Owing to the unusual peat ecosystem, highly acidic, anaerobic, and sensitive to fire and other weather conditions, earthworms are nearly absent from disturbed peat. Termites are therefore seen as a potential major soil engineer—if their services could be harnessed efficiently. To ensure the survival and growth of planted trees, farmers can take into account species selection and wood resistance to termite attacks. Otherwise, sustainable pest management and preventive measures should be devised before termites become a serious threat to newly planted trees. Local knowledge can harness termite power as an engine of resilience in degraded peatlands.

Chapter 7 by Suzuki discusses timber processing and retailing in Pekanbaru, Riau Province, Indonesia. Timber supply has changed along with its sources—from the peat swamp forest to the village forest and acacia plantation. The timber processing industry is quite flexible in this way. From hardwood timber to village forest timber such as Mahang, and Acacia, the source tree species have changed. Some molding mills and timber kiosks cope, depending on the pulp and paper companies. This is because it is easy to process and retail timber under company management, without any permission from local governments. On the other hand, the other molding mills and timber kiosks sell timber to local people and the government. They transform their management to adapt to the current situation of timber shortage, such as changing timber species from hardwood to softwood, selling any kind of lumber to retain timber size, and so on. This flexibility in the timber processing industry and retail businesses can be considered adaptability, and it might very well support the timber supplied by future paludiculture in the peat swamp restoration forest.

Chapter 8 by Gunawan et al. discuss the biomass and carbon restoration of the peat swamp ecosystem. It compares biomass and carbon storage among the peat swamp forest, exploited forest, and wind-disturbed forest. The amounts of biomass and carbon storage were higher in the peat swamp forest, followed by logged forest and wind-disturbed forest. Of the peat swamp forest tree species, growth rate of

Cratoxylum arborescens was highest, and followed by *Tetramerista glabra*. Some typical canopy species of peat swamp forest such as *Tetrameristra glabra* and *Palaquium burckii* are promising in their use to rehabilitate logged-over forest areas. These data show the possibility of reforestation with peat swamp forest tree species. On the other hand, carbon storage underground was quite high among any kind of peatland, natural forest, logged-over forest, wind-disturbed forest, oil palm forest, acacia forest, and rubber garden. In this case, it is important to avoid fires that could release carbon that is stored even in degraded peatlands.

The next four papers consider transformation of the peatland ecosystem. Chapter 9 by Sutikno et al. discuss water management for integrated peatland restoration. This study of water management was conducted at the Pulau Tebing Tinggi Peatland Hydrological Unit (PHU) with the purpose of peat rewetting and supporting revegetation efforts and revitalization of livelihood. The canal block constructions, paludicultures, and aquacultures were the integrated activities conducted to support peatland restoration. Two types of canal blocks constructed from wood and vinyl sheet pile were introduced in this pilot project. Four key parameters indicating progress in peatland restoration were monitored periodically, including water table, land subsidence, emission, and vegetation growth.

Chapter 10 by Tata presents the case for plant genetic diversity in peatland restoration. To achieve the target of peatland restoration with species survival rate of 90%, there must be a sufficient quantity of quality planting stocks to ensure genetic value and plant health. The strategy should involve various seed combinations, including not only local species but also those native to other similar environments. The planted jelutung in Jambi and Tumbang Nusa, Central Kalimantan, showed no loss of genetic diversity. Plants in those areas could be used as a source of seeds for peatland restoration throughout both provinces. Villagers with seedling nursery businesses should be made aware of the potential risk of genetic drift if they are collecting seeds from a few individual trees and limited populations. Strategies of seed sourcing should be introduced to farmers so that they are aware of the importance of genetic diversity in peatland landscape restoration.

Chapter 11 by Widyatmoko discusses Indonesian Sustainable Palm Oil Certification (ISPO). There are many ISPO holders among companies, but quite a few are found in small farmer communities, as small farmers are also encouraged to get the certificate to promote sustainable palm oil production. This study points out the difficulties that farmers face in getting the certificate and their various efforts to overcome these challenges, such as the support of local governments, companies, and community initiatives. Obtaining land title is of paramount concern, as are land conflicts that spring from questions of the legitimacy of the legal documents. The forest/timber certificate is among the important issues in peatland transformation (Carlson et al. 2018; Ehrenberg-Azcárate and Peña-Claros 2020). This study also examines the process of ISPO acquisition by small farmers in Riau.

As a whole this book hopes to shed light on the developments in peat swamp rehabilitation, and the way they contributed to transformation and regeneration of peatland landscapes through the integrated lens of vulnerability-resilience-adaptability-transformation.

Future volumes in this series will discuss other important aspects of peatland socio-ecosystems—governance, public health issues related to haze, the history and social perspectives of peatland societies, the impacts and implications of meteorological phenomena, and peatlands as they relate to the national and international economies. These discussions will be related to the shifts in perception and meaning, social network configurations, patterns of interaction among actors, and associated organizational and institutional arrangement. The authors look forward to publishing their findings on these aspects soon.

Here we would like to express our sincere gratitude to Springer and RIHN, the Research Institute for Humanity and Nature, which handled the publication of this book, and the JSPS (Japan Society for the Promotion of Science), which supports us through the project of JSPS Grants-in-Aid for Scientific Research, Project No. 19H04350.

References

Agus F, Wahyunto, Dariah A et al (2010) Carbon budget and management strategies for conserving carbon in peatland: case study in Kubu Raya and Pontianak districts, West Kalimantan, Indonesia. In: Chen ZS, Agus F (eds) Proceedings of international workshop on evaluation and sustainable management of soil carbon sequestration in Asian countries, Bogor, September 2010. Indonesian Soil Research Institute, Bogor; Food and Fertilizer Technology Center for the Asian and Pacific Region, Taipei. National Institute for Agro-Environmental Sciences, Tsukuba

Anderson JAR, Muller J (1975) Palynological study of a Holocene peat and a Miocene coal deposit from NW Borneo. Rev Palaeobot Palynol 19(4):291–351

Andriesse JP (1988) Nature and management of tropical peat soils. FAO Soils Bulletin 59. FAO, Rome

Anshari G, Kershaw AP, van der Kaars S (2001) A Late Pleistocene and Holocene pollen and charcoal record from peat swamp forest, Lake Sentarum Wildlife Reserve, West Kalimantan, Indonesia. Palaeogeogr Palaeoclimatol Palaeoecol 171(3–4):213–228

Anshari G, Kershaw AP, van der Kaars S et al (2004) Environmental change and peatland forest dynamics in the Lake Sentarum area, West Kalimantan, Indonesia. J Quat Sci 19(7):637–655

Applegate G (2012) The impact of drainage and degradation on tropical peatland hydrology, and its implications for effective rehabilitation. Paper presented at the international peat society 14th international peat congress on peatlands in balance, Stockholm, 3–8 June 2012

ASEAN (2007) Rehabilitation and sustainable use of peatlands in South East Asia. International Fund for Agricultural Development and Global Environment Facility, Rome

Carlson KM, Heilmayr R, Gibbs HK et al (2018) Effect of oil palm sustainability certification on deforestation and fire in Indonesia. PNAS 115(1):121–126. https://doi.org/10.1073/pnas.1704728114

Darusman T, Lestari DP, Arritadi D (2021) Management practice and restoration of the peat swamp forest in Katingan-Mentaya, Indonesia. In: Osaki M, Tsuji N, Foead N et al (eds) Tropical peatland eco-management. Springer, Singapore, pp 381–409. https://doi.org/10.1007/978-981-33-4654-3_13

Dennis C, Aldhous P (2004) Biodiversity: a tragedy with many players. Nature 430:396–398. https://doi.org/10.1038/430396a

Dohong A, Abdul Aziz A, Dargusch P (2018) A review of techniques for effective tropical peatland restoration. Wetlands 38(2):275–292. https://doi.org/10.1007/s13157-018-1017-6

Ehara H, Toyoda Y, Johnson DV (eds) (2018) Sago palm: multiple contributions to food security and sustainable livelihoods. Springer, Singapore. https://doi.org/10.1007/978-981-10-5269-9

Ehrenberg-Azcárate F, Peña-Claros M (2020) Twenty years of forest management certification in the tropics: major trends through time and among continents. For Policy Econ 111:102050. https://doi.org/10.1016/j.forpol.2019.102050

Evers S, Yule CM, Padfield R et al (2016) Keep wetlands wet: the myth of sustainable development of tropical peatlands – implications for policies and management. Glob Change Biol 23(2): 534–549. https://doi.org/10.1111/gcb.13422

Field RD, van der Werf GR, Shen SSP (2009) Human amplification of drought-induced biomass burning in Indonesia since 1960. Nat Geosci 2:185–188. https://doi.org/10.1038/ngeo443

Folke C, Chapin FS, Olsson P (2009) Transformations in ecosystem stewardship. In: Folke C, Kofinas GP, Chapin FS (eds) Principles of ecosystem stewardship: resilience-based natural resource management in a changing world. Springer, New York, pp 103–125. https://doi.org/10. 1007/978-0-387-73033-2_5

Folke C, Carpenter SR, Walker B et al (2010) Resilience thinking: integrating resilience, adaptability and transformability. Ecol Soc 15(4):20

Furukawa H (1992) Indoneshia no teishicchi. Keiso Shobo, Tokyo [English edition: Furukawa H (1994) Coastal wetlands of Indonesia: environment, subsistence and exploitation (trans: Hawkes PW). Kyoto University Press, Kyoto]

Huitema D, Meijerink S (eds) (2009) Water policy entrepreneurs: a research companion to water transitions around the globe. Edward Elgar, Cheltenham

Ismail Z (1999) Survey of fish diversity in peat swamp forest. In: Yuan CT, Havmoller P (eds) Sustainable management of peat swamp forest in Peninsular Malaysia. Forestry Department Peninsular Malaysia, Kuala Lumpur, pp 173–198

Jaenicke J, Rieley JO, Mott C et al (2008) Determination of the amount of carbon stored in Indonesian peatlands. Geoderma 147(3–4):151–158. https://doi.org/10.1016/j.geoderma.2008. 08.008

Jaenicke J, Wösten H, Budiman A et al (2010) Planning hydrological restoration of peatlands in Indonesia to mitigate carbon dioxide emissions. Mitig Adapt Strateg Glob Change 15(3): 223–239. https://doi.org/10.1007/s11027-010-9214-5

Jelles JGG (1929) De Boschexploitatie in het Panglonggebied. Tectona deel VXXII:482–506

Kato T, Silsigia S, Yusup AA et al (2021) Large-scale practice on tropical peatland eco-management. In: Osaki M, Tsuji N, Foead N et al (eds) Tropical peatland eco-management. Springer, Singapore, pp 89–134. https://doi.org/10.1007/978-981-33-4654-3_3

Kieft J, Smith T, Someshwar S et al (2016) Towards anticipatory management of peat fires to enhance local resilience and reduce natural capital depletion. In: Renaud FG, Sudmeier-Rieux K, Estrella M et al (eds) Ecosystem-based disaster risk reduction and adaptation in practice. Advances in natural and technological hazards research, vol 42. Springer, Cham, pp 361–377

Marlier ME, Liu T, Yu K et al (2019) Fires, smoke exposure, and public health: an integrative framework to maximize health benefits from peatland restoration. GeoHealth 3(7):178–189. https://doi.org/10.1029/2019GH000191

Mizuno K, Fujita SM, Kawai S (eds) (2016) Catastrophe and regeneration in Indonesia's peatlands: ecology, economy and society. Kyoto CSEAS series on Asian Studies, vol 15. NU S Press/ Kyoto University Press, Singapore/Kyoto

Mizuno K, Hosobuchi M, Ratri DAR (2021) Land tenure on peatland: a source of insecurity and degradation in Riau, Sumatra. In: Osaki M, Tsuji N, Foead N et al (eds) Tropical peatland eco-management. Springer, Singapore, pp 627–649

Momose K (2002) Environments and people of Sumatran peat swamp forests II: distribution of villages and interactions between people and forests. Southeast Asian Stud 40(1):87–108. https://doi.org/10.20495/tak.40.1_87

Momose K, Shimamura T (2002) Environments and people of Sumatran peat swamp forests I: distribution and typology of vegetation. Southeast Asian Stud 40(1):74–86. https://doi.org/10. 20495/tak.40.1_74

Murdiyarso D, Hergoualc'h K, Verchot LV (2010) Opportunities for reducing greenhouse gas emissions in tropical peatlands. PNAS 107(46):19655–19660. https://doi.org/10.1073/pnas. 0911966107

Olsson P, Folke C, Hahn T (2004) Social–ecological transformation for ecosystem management: the development of adaptive co-management of a wetland landscape in southern Sweden. Ecol Soc 9(4):2

Page SE, Hooijer A (2016) In the line of fire: the peatlands of Southeast Asia. Philos Trans R Soc B 371(1696):20150176. https://doi.org/10.1098/rstb.2015.0176

Page SE, Rieley JO (1998) Tropical peatlands: a review of their natural resource functions, with particular reference to Southeast Asia. Int Peat J 8:95–106

Page SE, Siegert F, Rieley JO et al (2002) The amount of carbon released from peat and forest fires in Indonesia during 1997. Nature 420:61–65. https://doi.org/10.1038/nature01131

Page SE, Rieley JO, Wüst R (2006) Lowland tropical peatlands of Southeast Asia. In: Martini IP, Martínez Cortizas A, Chesworth W (eds) Peatlands: evolution and records of environmental and climate changes. Developments in earth surface processes, vol 9. Elsevier, Amsterdam, pp 145–172. https://doi.org/10.1016/S0928-2025(06)09007-9

Page SE, Hosciło A, Wösten H et al (2009) Restoration ecology of lowland tropical peatlands in Southeast Asia: current knowledge and future research directions. Ecosyst 12(6):888–905. https://doi.org/10.1007/s10021-008-9216-2

Pastor G (1927) De Panglongs. Landsdrukkerij, Weltevreden

Posa MRC, Wijedasa LS, Corlett RT et al (2011) Biodiversity and conservation of tropical peat swamp forests. Bioscience 61(1):49–57. https://doi.org/10.1525/bio.2011.61.1.10

Purnomo H, Shantiko B, Sitorus S et al (2017) Fire economy and actor network of forest and land fires in Indonesia. For Policy Econ 78:21–31. https://doi.org/10.1016/j.forpol.2017.01.001

Purnomo H, Okarda B, Shantiko B et al (2019) Forest and land fires, toxic haze and local politics in Indonesia. Int For Rev 21(4):486–500. https://doi.org/10.1505/146554819827906799

Reep C (1907) Het eiland Bengkalis. Tijdschrift voor het Binnenlands Bestuur 32:375–398

Ritzema H, Limin S, Kusin K et al (2014) Canal blocking strategies for hydrological restoration of degraded tropical peatlands in Central Kalimantan, Indonesia. Catena 114:11–20. https://doi. org/10.1016/j.catena.2013.10.009

Saputra E (2019) Beyond fires and deforestation: tackling land subsidence in peatland areas, a case study from Riau, Indonesia. Land 8(5):76. https://doi.org/10.3390/land8050076

Sewandono M (1937) Inventarissatie en inrichting van de veenmoerasbosschen in het panglonggbied van Sumatra's Oostkust. Tectona XXX:660–679

Smith A, Stirling A (2010) The politics of social-ecological resilience and sustainable socio-technical transitions. Ecol Soc 15(1):11

Tata HL (2019) Mixed farming systems on peatlands in Jambi and Central Kalimantan provinces, Indonesia: should they be described as paludiculture? Mires Peat 25:08. https://doi.org/10. 19189/MaP.2018.KHR.360

Thornton SA, Setiana E, Yoyo K et al (2020) Towards biocultural approaches to peatland conservation: the case for fish and livelihood in Indonesia. Environ Sci Pol 114:341–351. https://doi. org/10.1016/j.envsci.2020.08.018

Uda SK, Hein L, Sumarga E (2017) Towards sustainable management of Indonesian tropical peatlands. Wetl Ecol Manag 25(6):683–701. https://doi.org/10.1007/s11273-017-9544-0

Uda SK, Hein L, Atmoko D (2019) Assessing the health impacts of peatland fires: a case study for Central Kalimantan, Indonesia. Environ Sci Pollut Res 26(30):31315–31327. https://doi.org/10. 1007/s11356-019-06264-x

Uda SK, Hein L, Adventa A (2020) Towards better use of Indonesian peatlands with paludiculture and low-drainage food crops. Wetl Ecol Manag 28(3):509–526. https://doi.org/10.1007/s11273-020-09728-x

van Beukering P, Schaafsma M, Davies O et al (2008) The economic value of peatland resources within the Central Kalimantan peatland project in Indonesia: perceptions of local communities. Institute for Environmental Studies, Vrije Universiteit, Amsterdam

Varkkey H (2013) Oil palm plantations and transboundary haze: patronage networks and land licensing in Indonesia's peatlands. Wetlands 33(4):679–690. https://doi.org/10.1007/s13157-013-0423-z

Walker B, Holling CS, Carpenter SR et al (2004) Resilience, adaptability and transformability in social–ecological systems. Ecol Soc 9(2):5

Wösten JHM, Clymans E, Page SE et al (2008) Peat-water interrelationships in a tropical peatland ecosystem in Southeast Asia. Catena 73(2):212–224. https://doi.org/10.1016/j.catena.2007.07.010

Part I
Vulnerability of Peat Swamp Forest

Chapter 2
Peatland Degradation, Timber Plantations, and Land Titles in Sumatra

Kosuke Mizuno, Kazuya Masuda, and Almasdi Syahza

Abstract Peatlands in Riau, Sumatra were relatively untouched by development or deforestation until at least the beginning of the 1970s. But today these landscapes are seriously degraded, with fires breaking out almost every year. Why and how has it come to this? This study attempts to make clear the relationships between the establishment of timber plantation, construction of large-scale drainage infrastructure, peatland degradation, in-migration, increasing fire events, and abandonment of peatland. This study highlights land rights as a factor that may either promote peatland degradation or motivate local people to manage degraded peatlands to better ends. It shows how large-scale drainage introduced by timber plantations since 1990s led to peatland desiccation in Riau, leading to fire events outside the plantation concession areas. Local people reacted to fire by logging and distributing parcels of peatland swamp forest outside the concession to secure land rights and to stop further concession giving to companies by the government. These activities in turn promoted peatland degradation, increasing the incidence of fire and abandonment of peatland. Local people's scramble to secure land rights promoted peatland degradation, but as soon as they obtained land titles they managed the burned lands well. On the other hand, land distributed land without title tended to be abandoned after fires. The intrusion of timber plantations and land distribution also promoted in-migration, which contributed to peatland degradation. One of the reasons why people could distribute these peat swamp forests among themselves was poor governmental management of state forest lands, as the boundaries between the state and nonstate forests remained unclear, especially for the local people.

K. Mizuno (✉)
School of Environmental Science, University of Indonesia, Jakarta, Indonesia

Center for Southeast Asian Studies, Kyoto University, Kyoto, Japan
e-mail: mizuno@cseas.kyoto-u.ac.jp

K. Masuda
Faculty of Agriculture and Marine Science, Kochi University, Nankoku-shi, Kochi, Japan

A. Syahza
Institute of Research and Community Services, University of Riau, Pekanbaru, Riau, Indonesia

© The Author(s) 2023
K. Mizuno et al. (eds.), *Vulnerability and Transformation of Indonesian Peatlands*,
Global Environmental Studies, https://doi.org/10.1007/978-981-99-0906-3_2

Keywords Peatlands · Degradation · Fire · Abandonment · Land title · Sumatra · Acacia plantation · State forest

2.1 Introduction

Peatlands in Riau, Sumatra were covered with dense forest at least until the beginning of the 1970s. Today this area is experiencing serious land degradation, with fires breaking out almost every year. Why and how has it come to this? Many studies on peatland degradation have identified key factors contributing to forest degradation. These include the establishment of timber and oil palm plantations, large-scale drainage projects, intensive land use associated with increased in-migration, and fire. This study picks up from previous research, emphasizing land rights as another important factor that may either promote peatland degradation or motivate local people to better manage degraded peatland.

This chapter aims to show three points based on fieldwork conducted in Bengkalis district, Riau Province, Sumatra, Indonesia between 2010 and 2021. First, it describes historical change in peatland use in the study area. Second, it describes the process of establishing land tenure among local people. Third, it discusses the relationship between the characteristics of land rights and peatland degradation.

2.1.1 Background: Development Programs and Peatland Degradation

Indonesian peat swamp forests were relatively untouched by modern state development initiatives until the early 1960s (Silvius and Suryadiputra 2005). Although the Riau area in Sumatra was logged commercially under the *Panglong* system since the 1860s, there were no related reports of peatland degradation or large-scale fires (Pastor 1927; Jelles 1929; Sewandono 1937). The peat swamp forests have been inhabited by Malay people for a long time with only limited or localized associated ecological damage (Page et al. 2009, pp. 900–01; Furukawa 1992; Momose 2002).

Several important changes occurred in the 1970s, however, including the initiation of large-scale logging, a governmental transmigration program, and spontaneous migration to the peatland area. These changes mark the beginning of peatland degradation (Dohong et al. 2017). The experience of Riau is not unique, as assessments of peatlands throughout insular Southeast Asia reveal dramatic reduction in peat swamp forest cover since 1985 (Hooijer et al. 2006; Miettinen and Liew 2010; Fuller et al. 2011; Miettinen et al. 2012b). Conversion of peat swamp forests to industrial timber and oil palm plantations is often seen as one of the major causes of deforestation (SarVision 2011; Miettinen et al. 2012a). In particular, oil palm cultivation has caused much controversy (Stone 2007; Venter et al. 2008; Sheil et al. 2009). A majority (62%) of the plantations were located on the island of Sumatra, and over two-thirds (69%) of all industrial plantations were developed for oil palm cultivation, with the remainder mostly being acacia plantations developed for production of paper pulp. Historical analysis shows strong acceleration of plantation development in recent decades: 70% of all industrial plantations have been established since 2000, and only 4% of the current plantation area existed as plantation in 1990 (Miettinen et al. 2012a).

These developments led to systematic drainage of peat swamp forests which has triggered the long-term degradation of peatlands, increased carbon emission in this area (Dohong et al. 2017), and led to regular occurrence of fire. A survey in 2013–2014 (Gaveau et al. 2016) showed that of 404,713 ha burned in those years, 84% (330,000 ha) was peatland, 10% (38,451 ha) of the total burned area was mature acacia tree stands, and 18% (54,870 ha) of burned non-forest lands were oil palm stands. However, 75% (300,000 ha) of the burned area was previously non-forest land, and of this non-forest land, 82% was idle land, that is, unplanted peatlands covered with shrubs and wood debris.

How has the development of timber and oil palm cultivation been related to peatland degradation, and especially peatland fires, outside the concession areas? The idle lands identified by Gaveau et al. (2016) that are prone to burning probably refer to abandoned land. A village in Central Kalimantan surveyed by Maimunah et al. (2018) was affected by forest and peatland fires that destroyed large areas of productive agricultural land. Most of the burned land was then abandoned due to declining fertility, and the farmers began to look for alternative land uses to meet their livelihood needs (Carlson et al. 2013; Maimunah et al. 2018).

Joosten et al. (2012) reported that millions of hectares of the world's drained peatlands have such low productivity and have become so degraded that they have been abandoned. In the absence of management, abandoned and drained peatland sites are particularly susceptible to fire, as partially logged and previously burned forests in the tropics may accumulate considerable dead wood litter, and the dry peat beneath is easily ignited in the dry season.

Despite these accounts it is not clear how acacia and oil palm plantations and associated drainage projects degrade peatlands, or increase vulnerability to fire both within and without concession areas. Does peatland degradation generally lead to abandonment? How do local people who have lived with peatland resources respond to similar threats and challenges?

Studies on peatland degradation conducted by the Ex Mega Rice Project, or MRP (1995–1999), in Central Kalimantan province address these questions at least in part. MRP's goal was to turn 1 million ha of unproductive and thinly populated peat swamp forest into rice paddies in order to address Indonesia's food shortage. The project called for large-scale transmigrations, irrigation canals, and clearing vast areas of peat swamp forests. The project was eventually abandoned, but only after it had caused significant environmental damage and fire, and induced a livelihood crisis (Galudra et al. 2011; Jewitt et al. 2014). The construction of logging roads, land clearing, and a shift by local communities toward shorter fallows within farming systems led to large areas of deforestation, forest degradation, and large carbon emissions form the tropical peatlands of Central Kalimantan (Medrilzam et al. 2017; Chokkalingam et al. 2005; Law et al. 2015; Medrilzam et al. 2014).

Jewitt et al. (2014) described the history of local livelihood strategies in the area before the MRP, and compared them to conditions during the MRP, and as they were affected by programs following the MRP. Examining a community-based forest management program, logging concession given by the local government, and REDD+ Program in the former MRP area, they demonstrated the limited ability of government programs to rehabilitate degraded peatlands.

How does peatland degradation relate to acacia and oil palm plantations outside the MRP? What is the relationship between the introduction of acacia plantation, oil palm cultivation, drainage projects, in-migration, peatland degradation surrounding acacia plantations, fires, and land abandonment in those areas? How did local people, especially the Malay people, respond to changes? The first aim of this study is to answer these questions based on case study in Riau Province.

2.1.2 Land Ownership and Land-Use Management

This study examines how changes in land tenure relate to shifts in peatland use and degradation. Land tenure conditions influence the continuity and productivity of agricultural production, with many studies emphasizing the role of land-tenure and land-use rights as prerequisites to better land management by small holders (Suyanto et al. 2005). Feder and Noronha (1987) and Feder and Feeny (1993) strongly argue that secure private land ownership effectively incentivizes small farmers to invest in land improvement. Even if land tenure is not guaranteed by government title but is instead based on informal community institutions, it appears that secure access to land encourages farmers to adopt better land-use management.

Does this finding hold true in the case of peatlands? There are some studies on the role of secure land rights, both individual and communal, in promoting forestation or peatland rehabilitation programs or better land use. Nevertheless, there is persistent argument that securing land rights would promote further peatland degradation, especially suggesting that more people would enter the peatland forest in search of secure private land. How are land tenure, especially secure land rights, related to peatland degradation, or conservation and rehabilitation?

Wildayana et al. (2019) argue on the contrary that increasing secure land ownership would promote peatlands degradation. They suggested that granting tenure would empower farmers to claim future land rent and increase land clearing as farmers seek additional income, but the authors do not show any data or other information to support these arguments.

On the other hand, Jewitt et al. (2014) show the importance of customary land rights or land ownership in relation to peatland restoration programs such as REDD+ and community-based forestry management programs. Nawir et al. (2007) also show the significance of customary land ownership to community-based forestry management. Many papers have discussed land tenure in state forest lands (or the

Government-designated state forest area [*Kawasan Hutan*], hereafter refer to state forest, or the Government-designated state forest area, or state forest area, or state forest lands), including their relationship to land-use conflicts in the state forest (Yusran et al. 2017; Kunz et al. 2017); the agrarian reform program, especially the social forest program in state forest area in the peatlands (Resosudarmo et al. 2019); or land ownership and community involvement as a key factor for the success of the agroforestry program in state forest area (Suyanto et al. 2005). Meanwhile, Rietberg and Hospes (2018) described a case in which land acquisitions in an oil palm frontier obscured customary rights and local authority, and resulted in land conflict between the local people and the estate company.

Mizuno et al. (2021) showed that majority of peatlands in the study's research site are located in the state forest, and that there are overlapping land rights in the peatland area. Many people obtained land by clearing, inheritance, purchase, and distribution, and most of these transactions were conducted according to customary practice. Land rights based on customary practice, however, are frequently not recognized by the local government. This study extends Mizuno et al. (2021), examining the relationships between the introduction of acacia plantations and the prevalence of oil palm cultivation, land tenure and land rights, and peatland degradation or conservation. It examines households holding certificates of land rights or land titles, whether customary or statutory, or secured or non-secured, seeking to answer the question of whether secured land rights promote better land management, or encourage further peatland degradation. The particular status of land rights can be verified by examining the certifying land title letters. Many the villagers have letters of SKT (*Surat Keterangan Tanah*), which is based on customary rights, while other villagers have certificate letters of land ownership *(Sertifikat Tanah)* issued by the Indonesian National Land Agency (*Badan Pertanahan Nasional*, BPN). Some villagers have no letter to certify their land rights. Generally, land ownership letters are thought provide more secure land rights than SKT customary rights-based letters. Having a SKT letter is still thought to be better than having no letter at all, however.

2.2 Methodology

2.2.1 Selection of Research Site and Its General Description

Research was conducted in Tanjung Leban village (*Desa* Tanjung Leban), Bukit Batu sub-district (*kecamatan*), Bengkalis district (*kabupaten*), in the east part of Riau province (*propinsi*). This village is located in the peatland area that faces the Malacca Straits. The village was selected for the following reason.

Furukawa (1992) identified two different groups of people making use of peat swamp forests. Malays are characterized as members of a "culture of transit", people whose multiple livelihoods strategies shift as easily as do their domiciles. Traditionally, the Malays have not exploited the peatland intensively; if they have made use of it, they have done so without degrading the landscape—for example, by engaging in fishery, rubber production, small-scale slash and burn, and so on. The second group comprises immigrants such as Javanese and Banjarese who have exploited the peatland tidal forests on a large scale, clearing fields, planting rice, and establishing themselves as permanent residents (Furukawa 1992). This study selected the Malay village in order to better understand the process of peatland degradation in an area in which peatlands had been relatively well maintained at least until the beginning of the 1980s.

Tanjung Leban was originally a Malay settlement. This study traces the village history further back, as well as the in-migration to the village and expansion of the land exploited in the last few decades. The village significantly expanded at the end of the 1990s, partly because of the intrusion of timber plantations in the 1990s, and again in the 2000s due to the exploration of former peat swamp forests, the inflow of migrant workers in large-scale logging operations and timber smuggling, and oil palm cultivation. The village thus provides insights into the relationship between the intrusion of timber plantations and peatland degradation, the development of land use and tenure system in a traditional Malay village, and the expansion of oil palm cultivation in a former peat swamp forest, which is mainly the Government-designated state forest area.

Tanjung Leban village has a population of 1145 (601 male and 544 female), and where there were 321 households in 2010 (Bukit Batu sub-district 2010). The village covers 17,000 ha that extend to the protected area of the Giam Siak Kecil–Bukit Batu Biosphere Reserve. The bio-reserve was recognized by UNESCO in 2011 and consists of a 178,222-ha core area, a 222,425-ha buffer zone, and 304,123 ha of transition areas. Administratively, the bio-reserve belongs to both Bengkalis and Siak districts. Extensive *Acacia crassicarpa* and oil palm plantations are found in the buffer zone and transition areas. The surveyed village includes natural peat swamp forest, timber plantations, and areas owned or utilized by local people; most of these lands consist of peatland on a flat terrain. Only around 1 km from the seashore that faces the Malacca Straits is alluvial soil, and it is in this area, at the border between the alluvial soil and peatland, that people have traditionally settled. The peatland was covered by peat swamp forest a until the middle of the 1990s. A majority of this peat swamp forest is within the state forest area, and all of the Giam Siak Kecil Biosphere Reserve is within the state forest area and is therefore state land (Mizuno et al. 2016). According to the authors' calculations in April 2020 based on the map shown by the Peatland and Mangrove Restoration Agency (*Badan Restorasi Gambut dan Mangrove*, BRGM) of 27,960 ha of the research area, 24,358 ha (87.1%) was peatland, and 3602 ha (12.9%) was non-peatland; on the other hand, as much as 28,206 ha (99.1%) was state forest, and only 246 ha (0.9%) was APL (area of other use, outside of state forest area) (Mizuno et al. 2021).

Panglong logging, mainly led by Chinese entrepreneurs in Singapore and featuring the *ongka* logging system of wood rails and sleighs, has been conducted here since the 1920s (Jelles 1929, p. 484). The government has issued logging concessions (*Hak Pengusahaan Hutan*, HPH) since the 1970s and industrial forest plantation concessions (*Hutan Tanaman Industri*, HTI) to private companies in this state land since the 1990s. The landscape changed drastically in the 1990s as timber companies began planting *Acacia crassicarpa*, and trees grown in the concession areas began to be supplied to the giant paper company located at Perawang, Riau Province. Such timber plantations started operations at the end of the 1990s (Masuda et al. 2016).

Tanjung Leban consists of three major sub-villages (*dusun*), namely, Bakti, Air Raja, and Bukit Lemkung. Many migrants moved to Air Raja and Bukit Lemkung from North Sumatra since around 2000. The *Acacia crassicarpa* plantations and the accompanying road and canal construction made the thick peat swamp forest accessible to people from outside this region. The migrants to Air Raja and Bukit Lemkung are mainly Javanese born in the Medan area in North Sumatra.

In contrast, Bakti sub-village mainly comprises Malay people who have inhabited the area for a long time, principally on their own lands, some of which are not in the state forest. The authors chose to investigate Bakti sub-village because of the high percentage of Malays among the population and because it has the longest history among all the sub-villages of Tanjung Leban. It thus provides insight into customary law of the area as well as the changes that have taken place over time. Usually, a *dusun* comprises several hamlets. However, in the case of Bakti sub-village, there is only one hamlet (Bakti hamlet),[1] so the former is the same as the latter. For clarity, Bakti sub-village is hereafter referred to as the "surveyed sub-village" and Bakti hamlet as the "surveyed hamlet." The administrative village of Tanjung Leban is called the "surveyed village."

2.2.2 Data Collection

In total, the authors completed 71 household survey questionnaires in the Bakti hamlet. The primary survey was conducted from 2010 to 2012. Supplemental information about social and ecological changes was collected until the beginning of 2021. The authors investigated whether household lands are peatland or not; the history of land acquisition, land use, and land titling; peatland conditions, including the depth of the peat layer and whether and when the land has experienced fire; existing vegetation; and agricultural inputs and yields for each plot of land. They surveyed the composition of household members, education, occupation, ethnicity, birthplace, history of migration, and so on. Intensive study was concentrated on Bakti sub-village.

On the local perceptions of whether the land is peatland or not vary, each person provides his or her own answer. There are many lands that have a thin peat layer.

---

[1] The surveyed sub-village consists of two RWs and four RTs. An RT (*Rukun Tetangga*) is a neighborhood organization comprising around forty households. One RW (*Rukun Warga*) consists of two to four RTs. More Malay people live in RT01/RW01, RT02/RW01, and RT03/RW02 compared with RT04/RW02, where there are more migrants. Consequently, the authors conducted household surveys using a complete survey approach in RT03/RW02 and RT02/RW01. They also conducted household surveys to some extent in RW01/RW01.

Sometimes people say that the peat layer has already disappeared, or that "*menjadi tanah hitam,*" or "*menjadi kilang manis,*" i.e., "the land is not peatland anymore." The authors asked the peatland status of each plot of land. People sometimes answered that within the home garden, half is peatland and the other half is mineral soil. In these cases, the authors divided the plot and counted each part as either peatland, or non-peatland.

2.2.3 Land Tenure: de Jure Land Rights and de Facto Land Rights

The authors investigate land tenure and land title to determine the strength or security of land rights, and when and how those rights were obtained, and what relation they may have to peatland degradation. They mainly analyze the land *owned* by the respondents because sharecroppers or leasers do not know, or are not familiar with, the history of land acquisition, titling, and burning. As will be shown later, the amount of land leased to the respondents is small, so the authors' analysis is representative for the hamlet surveyed. All analyses compare the differences between peatland and non-peatland.

Land tenure is defined as "the right, whether defined in customary or statutory terms, that determines who can hold and use land, including forests and other landscapes and resources, for how long, and under what conditions" (Resosudarmo et al. 2014).

The discrepancy between de jure rights and de facto rights over land tends to manifest as land disputes, careless land use, or overlapping of land management subjects and rights. The conditions of de jure rights and de facto land rights can be studied in the process of land acquisition and land titling. For both of these there are customary and statutory practices. Customary land rights sometimes become the basis for statutory practices, but in most cases the state has tended to neglect customary way. There is a wide range of non-statutory land rights, from strongly secured to weak and unsecured. For example, secure customary rights are sometimes established through well-recognized letters or well-recognized custom, while in other cases some lands are claimed without any letter of certification or by obscure customary practices. This study pays special attention to such differences.

There are many different ways to certify land rights in Indonesia, and land prices differ according to the types of land titling (Mizuno and Shigetomi 1997). Land titles can secure the right to the land when the land is disputed, and titles issued by the agrarian office can be mortgaged. On the other hand, letters issued from the village office according to the customary way are not recognized by the agrarian office and cannot usually be mortgaged in banks.

During colonial times, land rights possessed by Europeans were based on the civil code (*Burgerlijk Wetboek*). Their lands were registered and a written certificate (*een schriftelijk bewijs*) was issued. Europeans could obtain right of land ownership (*eigendom recht*). On the other hand, Indonesians could get rights of possession (*bezitrecht*) on land based on customary law without registration. Indonesians did not then have land titles, only an excerpt from the land rent registration book (called Buku Letter C, and the excerpt was called *girik* or *kikitir,* among other names). This excerpt was considered a document demonstrating the amount of land tax owed, and identifying the payee, it was not considered a certification of land ownership. This excerpt was issued to people who possessed land and paid the land tax in the island of Java (Mizuno 1991), but not to those in islands outside Java and Madoera (Madura).

The Basic Agrarian Act (UUPA) of 1960 unified these dualistic systems of land rights, and stipulated that the right of land ownership would be based on customary law. The Act obliged the government to register land all over the country. However, Article 19 Clause 3 of the Act stipulated that land registration be implemented in consideration of the state of the government and society. Here, majority of the land owned by Indonesians began to exist as land with the right of ownership without registration or titling. Since the 1980s, the government implemented the National Land Registration Program (*Prona*), yet by 1991 the percentage of registered land only amounted to 1.9% of the total.[2] The program of land registration has proceeded at a somewhat similar pace since then.

In 2001, MPR (People's Representative Board) passed the "TAP MPR" a decision allowing the government to implement agrarian renewal (*Pembaruan Agraria*). The administration of current President Joko Widodo has targeted redistribution of 9 million ha through TORA (*Tanah Obyek Reforma Agraria*, agrarian reform program), which deals with land titling and redistribution to small-scale or landless farmers through Government Regulation No. 86 Year 2018 on Agrarian Reform (Resosudarmo et al. 2019; Muchsin et al. 2019; Arisaputra 2015). Land registration and land titling has therefore accelerated.

People whose land is registered posses a land certificate which varies depending on the kind of land right. The most common certificate letter is a land ownership certificate (*sertifikat hak milik tanah*). How were these enforced in the surveyed village?

In some places in Sumatra, a "letter of statement" had been issued to Indonesians who possessed land based on customary rights (the colonial system called this right *Inlander bezitrecht*, or the "possession right of indigenous people"). For example, in Jambi a regulation issued in 1930 (*Regeeringsomslagvel* No. 30318, 17 October 1930) enabled a village head (*kepala kampung*), or an assistant to the Resident, to make a letter of statement (*surat keterangan*) for an Indonesian who had inheritable rights of individual possession (*erflijk individueel bezitrecht*). The letter of statement was to be attached to a map (*schetskaart*). Letters of statement have been issued ever since. For example, from 1958 to 1 April 1963, the agrarian office in Jambi City and

---

[2] The area of land that had been registered was only 1.9% for all land in Indonesia. On the other hand, in areas that have been able to levy land tax outside the forest, 6.8% was registered at the beginning of the 1990s (Harahap 1991/1992, p. 26).

Jambi Region (*Kantor Agraria Daerah dan Kantor Agraria Kota*) issued thousands of letters of statement such as the Letter of Statement for the Right on Land (*Surat Keterangan Hak Tanah,* SKHT) and the Letter of Statement for the Land Ownership (*Surat Keterangan Hak Milik*, SKHK) (Parlindungan 1978, pp. 16–32).

Under the Basic Agrarian Act of 1960, Letters of Statement made by the village head or excerpts from the land rent registration book were incorporated into the agrarian office administrative system. If a person wishes to register land, they can submit a registration proposal with the letter of statement made by the village head and recognized by the assistant of the *wedana* (today the *camat,* or head of the sub-district), and the excerpt from the land rent registration book mentioned above. If the proposal is approved, this makes clear the right to the land according to Government Regulation No. 10 of 1961 regarding land registration.[3]

Recent government regulations tend to use the role of the village head more actively. Government Regulation No. 24 of 1997[4] on land registration stipulates that land without any supporting documents can be certified when the land is de facto managed or overseen by the person for 20 years or more, at which point a letter of statement can be issued by the village head.

The government tends not to recognize letters issued by village office or sub-village office (SKT discussed later) on the state forest (*kawasan hutan*), however. Local government has repeatedly warned the sub-district office and village office not to issue SKT in state forest lands, and the 1998 Forestry Act prohibited people from making use of state forest without permission from the Government (Mizuno et al. 2021).

How many people own such letters in the surveyed village/hamlet? It is entirely possible that other kinds of letters also exist. When registered land is transferred, the official transaction document depends on the kind of transaction. In the case of a land purchase, an official document of land purchase (*Akta Jual Beli Tanah*) should be made by the Land Official Documents Officer (*Petugas Pembuatan Akta Tanah*, PPAT) or the head of the sub-district (*camat*), and submitted to the agrarian office by the landholder. When the land is not yet registered, people who wish to have it registered should make the certificate of transaction according to Government Regulation No. 10 of 1961. Thus, people often rely on the certificate of land purchase to certify their right to the land. In some cases they have this document even before the land is registered (Mizuno 1991). Moreover, for land that is not registered, people sometimes have the official document of land purchase which certifies the right to the land.

---

[3] Peraturan Pemerintah No. 10 tahun 1961 tentang Pendaftaran Tanah. Recent government regulations, especially those issued by local governments, tend to use the SKT more actively. Government Regulation No. 24 of 1997 on land registration (Peraturan Pemerintah No. 24 Tahun 1978 tentang Pendaftaran Tanah) stipulates that land without any document to certify land rights can be certified when the land is controlled de facto by the person for 20 years or more, and letter of statement can be issued by the village head.

[4] Peraturan Pemerintah No. 24 Tahun 1997 tentang Pendaftaran Tanah.

How are these conditions reflected in the survey site? And how do these different titling documents and condition related to changes in land tenure in general, and to use of peatland and peatland degradation, including fire and land abandonment? The following section describes our findings.

2.3 Findings

2.3.1 Formation of the Surveyed Settlement, In-Migration, and Peatland Degradation

2.3.1.1 Changes in Peat Swamp Forests Were Closely Related to Changes in the Local Population's Ethnic Composition

Present-day respondents to our household surveys began living in the hamlet in the 1930s. One Malay respondent born in 1952 in the surveyed hamlet said that his father, who was born in Bengkalis Island, moved to the surveyed hamlet in the 1930s. Some other families moved to the hamlet during the Japanese occupation of 1942–1945.

Since that time permanent settlement (even as people remained mobile across the area) is relatively recent compared to the Bukit Batu area along the Bukit Batu River. According to an encyclopedia of geography published in 1869, "At Bukit Batu, there are three hundred to four hundred fishery boats that belong to the houses in this area. In the same name village along with the same name river there is a small fishery port" (Veth 1869, p. 173).[5] The surveyed hamlet is about 60 km from Sungai Pakning town, the capital of Bukit Batu sub-district. Sungai Pakning has a port from which a regular ferry boat departs for Bengkalis Island, where the capital of the district is located about 50 km from the surveyed hamlet.

Around 200 m south of the coast of the surveyed village, a main road connects Sungai Pakning and a big oil-port town, Dumai, which is located 30 km to the northwest. At the moment this is good paved road newly built in the 1990s. It replaced a small road (built in the 1970s) nearer the seashore that connected both towns. Before that there was no road, and so the people went to these towns on small sailboats, especially as needed to transport birthing mothers and the sick.

In the 1920s, Bengkalis Regency (Afdeeling Bengkalis) had a population density of just 2.23 persons/km^2, yet in 1925 it exported 6000 tons of rubber, making it the largest exporter of any of the sub-regencies on the east coast of Sumatra. There were 1.36 million rubber trees in Kampar-Siak district, which included the surveyed village at the time. Of these, latex was harvested from 1.05 million trees, and

---

[5]When Englishman Anderson made an expedition in 1823, he mentioned Bukit Batu as a place of considerable trade and large fish catches (3–400 boats with 2–3 persons each were found, especially for fisheries) (Schadee 1918, p. 39).

3200 tons of rubber were exported (Departement van Landbouw, Nijverheid, en Handel 1926, pp. 15–18, 28).

We can see from the above that rubber cultivation spread extensively throughout Bengkalis Regency as early as the 1920s. According to present-day residents, rubber has been cultivated at least since the 1930s, but widespread cultivation of rubber trees in the surveyed village occurred after the 1950s. On the other hand, betel palm trees were planted mainly in the surveyed hamlet until the 1950s.[6] *Panglong* logging extended operations in Bengkalis Regency in the 1920s, and was the most active business in Indonesia at the time. The most important logging activities then took place in the peat swamp forest areas (Endert 1932, p. 733).

All villagers who began living in the surveyed hamlet in the 1930s were Malay, and many of their children were born in the surveyed hamlet. Some of them married fellow villagers; however, many married couples began arriving from outside the hamlet. Several such couples that moved into the hamlet were Javanese, some were born in Riau, but many were immigrants from Java Island, or Javanese who came from North Sumatra, particularly the Medan area.

Here, if one member of a couple originated from the surveyed village, the couple is categorized as "local residents." In other words, a "migrant" couple is one in which both members of the couple are not from the hamlet. Ethnic makeup is determined by the ethnicity of the parents of each member of the couple. In the case of migrants, the year of arrival is the year the first member of the couple arrived in the village. In the case of local residents, the arrival year indicates that the couple's earlier year of birth in the surveyed hamlet.

Among 71 respondent households, 34 contained one or more members originally from the surveyed hamlet. Among those 34 households, 14 are Malay-Malay couples. The remaining couples are Malays married to someone of different ethnicity, with the exception of one Javanese couple. The majority of mixed-ethnicity spouses are Javanese (15 cases include couples of mixed ethnicity of Malay, Javanese, and others). Other ethnic groups in this table include Minangkabau, Banjare, Sundanese, Buginese, Chinese, and Ocu Bangkinan. The Ocu Bangkinan are a minority group that mainly stays in Kampar district, Riau, which is adjacent to the Minangkabau's area in West Sumatra.

In the remaining 37 couple-based households, neither member was born in the surveyed hamlet. Among these migrants, 17 households migrated to the surveyed hamlet prior to 1995, when there were no large-scale *Acacia crassicarpa* plantations (prior to 1984, 7 households comprised migrant couples; and from 1985–1994, 10 households had migrant couples).

The main economic activities in the surveyed village prior to 1995 were in fisheries, dry rice shifting cultivation (*ladang*), rubber cultivation, and logging. Vast peat swamp forests were found, yet people cultivated the non-peatland and the border areas between peatlands. Secondary forest areas increased due to logging

---

[6] Authors' interviews with respondents No. 16, 30, 60, and 61 in March 2011, October 2014, and December 2014.

(Watanabe et al. 2016). Yet the technical difficulties of the traditional system of *ongka*, which was used for logging prior to 1995, limited logging areas (Watanabe et al. 2016; Masuda et al. 2016; Momose 2002). In-migration also increased prior to 1995, partly because the surveyed village was considered a place with rich resources to be exploited, and partly because there were opportunities to work in logging—conducted by either the concession company (HPH) or local people. Until this time, land was acquired through inheritance, purchase, or clearing the forests, as will be described in the following section. Opening up of the road connecting Sungai Pakning and Dumai in the 1990s enabled further inflows of people. The matter of their land acquisitions will also be discussed in detail below.

Table 2.1 accordingly shows that the surveyed hamlet was formed first by Malays in the 1930s; they were gradually followed by an inflow of Javanese who married local residents during 1950–1980. Finally, migrant couples started to move into the village in significant numbers after 1985, and this increased after 1995.

2.3.1.2 Peatland Degradation in the Hamlet Is Closely Related to Intrusion of Timber Plantation Since 1998

The *Acacia crassicarpa* plantations were established in this area in 1998 by a timber company representative of the paper, pulp, and timber industry, while a number of companies received industrial tree plantation concessions in the beginning of the 2000s at the Giam Siak Kecil area. In order to plant *Acacia crassicarpa*, the groundwater table level should be reduced to 70 cm below the soil surface.[7] Many ditches were built to reduce the peatlands water table, discharging huge amounts of water into the sea. These ditches dried out the peat swamps, rendering them extremely vulnerable to fire. Fires began to appear in the dried peatland surrounding the acacia plantation at the end of the 1990s.

This period also marks the beginning of large-scale logging in the remaining peat swamp forests. After President Suharto stepped down in 1998, rules and laws were generally loosened in Indonesia, and traders from Malaysia became more willing to buy illegally logged timber. Informal village leaders organized logging groups equipped with chainsaws, vessels to carry timber, and heavy equipment such as power shovels, sometimes with financing provided by the timber traders. The groups would log a particular block of the forest, and then distribute the former forest lands to the villagers. Many of these distributed lands were part of the state forest area. For example, a block consisting of upward of 300 ha was logged at the initiative of informal local leaders who had the businesses such as timber trader backed

---

[7] Supiandi reports that *Acacia crasicarpa* and oil palm can grow at the underground water level of 60–100 cm at the peatland (Sabiham 2009, pp. 242–243).

Table 2.1 The ethnic makeup of surveyed household couples in 2010 and 2011, the origin of surveyed households, and the year when they took up residence in the surveyed hamlet (unit: number of households) (Source of data: Authors' survey in 2011–2014)

	Local residents					Migrants						Total
	Malay Malay	Malay Javanese Others[a]	Javanese Javanese	Malay Others[a]	Total	Malay Malay	Malay Javanese Others[a]	Javanese Javanese	Malay Others[a]	Javanese Others[a]	Total	Total
1935–1944	0	0	0	0	0	3	0	0	0	0	3	3
1945–1954	1	1	0	0	2	0	0	0	0	0	0	2
1955–1964	2	3	0	0	5	0	0	0	0	0	0	5
1965–1974	7	4	0	2	13	0	2	0	0	0	2	15
1975–1984	4	4	0	2	10	0	2	0	0	0	2	12
1985–1994	0	1	0	0	1	5	3	2	0	0	10	11
1995–2004	0	2	0	0	2	2	7	5	2	1	17	19
2005–2011	0	0	1	0	1	0	2	0	1	0	3	4
Total	14	15	1	4	34	10	16	7	3	1	37	71

[a]Others are Minangkabau, Sudanese, Buginese, Chinese, Banjar, and Ocu Bangkinang

financially by Malaysian traders. The leaders received half of the lands cleared, while the remaining part was distributed to the nearby villagers in parcels of about 2 ha each.[8]

An informant reported that local people also advanced to clear the peat swamp forest in order to counter the further intrusion of industrial timber plantations into lands they viewed as their own lands.[9] In this way, deforestation was a strategy by local people to secure their traditional their common lands in the peat swamp forest. Such small-holders would also plant oil palm on cleared peatlands, and since oil palm cannot grow in peat swamp conditions, people intentionally kept the lands dry, making the peatland even more vulnerable to fire. Many ditches were expanded in the period of large-scale illegal logging between 1998–2006, further drying the peat swamp.

Watanabe et al. (2016) analyzed land-use changes in the 21,800 ha around our surveyed village of Tanjung Leban. In 1993, 77.5% of the area was natural forest; secondary forest covered 14.1%; and only 5.1% was farmland or oil palm land. However, with the start of acacia plantations in 1998, large-scale logging and subsequent opening up of farmland for oil palm, the area of natural forest decreased to 29.5% by 2006. Area of Acacia plantations increased to 37.2%, and farmland or oil palm land increased to 20.1% by 2006.

This drastic change in land use, the influx of new businesses related to the large-scale logging, distribution of former forest lands and new possibility of small-scale cultivation of oil palm: all attracted many immigrants. These migrants were usually first engaged in many sorts of casual labor, including oil palm maintenance and management of small shops. Later they obtained parcels of peatland and began to plant oil palm. At the same time, the original Malay settlers also often planted oil palm in the newly distributed lands. The change in landscape has altered the attitude of traditional Malays who once made use of the land without transforming the landscape drastically.

The influx of immigrants also changed the ethnic composition of the surveyed village from pure Malay to a complex composition of Malay and others, especially Javanese. This shows the patterns of migration and settlement through marriage and the openness of Malay people to other ethnic people.[10] Today the head of the surveyed village is always a Malay, but Javanese also occupy some important positions such as head of the Village Consultation Body (*Badan Persyawataran Desa*, BPD).

Watanabe et al. (2016) shows that in 2010, 18% of the area around the village was covered in immature oil palms; 4.7% of land was covered in mature oil palms.

---

[8] Authors' interview with respondents such as No. 16 on December 16, 2014, No.73 on December 14, 2017

[9] Authors' interview with respondent No 6 on January 22, 2022.

[10] In the surveyed hamlet, there are Christian Batak people who keep pigs. There are Chinese rubber merchants. Majority of the people there are Muslim. This information shows the openness of Malay people. These conditions cannot be imagined at the Sundanese village where author had conducted intensive field research in Bandung district, West Java (Mizuno 1996).

Although many lands were planted with oil palm, the palms often require 3 years before they are productive, and in this time, they require relatively dry soils. Many oil palm fields therefore burned before reaching maturity, and those lands became barren or turned into grasslands. In 2010, barren land covered 18.8% of the area around the surveyed village; grasslands covered 18.5%. As Gaveau et al. (2016) deduced, draining in concession lands promotes fire outside the concession. Concession companies typically have facilities to prevent fire, such lookout towers and fire brigades, however, so were able to protect their own assets from fires, while small-holders had little ability to prevent fires on their lands.

The following section will discuss changes in land use and land tenure along with peatland degradation. The authors investigate land tenure and type of land use, how the land was acquired, and the documents certifying land titles.

2.3.1.3 Change in Land Tenure Along with Peatland Degradation and In-Migration

The 71 households in the surveyed hamlet own a total of 843.2 ha, with each household owning an average of 11.9 ha. Some of their lands are located outside of the hamlet and even outside of the boundary of the surveyed village. Peatlands account for 660.9 ha or 9.3 ha of households landholdings. Cultivated—or operated—land accounts for 536.4 ha, or an average of 7.6 ha per household. The surveyed households lease a total of 20.2 ha (an average of 0.3 ha) while people from outside the hamlet lease a total of 9.3 ha. The gap between the amount of land owned and the amount of land operated is due to the large area that is not being operated at all. Most uncultivated land is abandoned; it comprises 264.7 ha of the total area, or 3.7 ha per household. A total of 33.0 ha is reserved land; households reserve land when they haven't yet decided on how to use it, when it's fallow, or if they have more land than they need at any given moment. Nearly 83.8% of owned land—a total of 690.2 ha—has been burned, the majority of it peatland.

The short history of the surveyed hamlet described above partly indicates the process of land acquisition. Here the authors explain the process more systematically for all land owned by respondents, and its relation to peatland vs non-peatland, and finally, to land titles.

Inheritance (*warisan*), donation during lifetime (*bagi-bagikan tanah kepada anak* or *hibah*), purchase (*beli*), and clearing the forest to convert to farmland (*buka lahan sendiri*) are the main means of acquiring land. As explained earlier, the distribution of land (*pembagian lahan oleh masyarakat*) has recently picked up pace.

Clearing the forest to make farmlands is customary in the village. Some respondents in the fields said that they paid *uang pancang* (*uang* means money, *pancang* means the place marked with stake at the hillslope to indicate a person's intention to cultivate the land [Adatrechtbundel 1916, pp. 174, 187]), or money indicating intention to cultivate the land) to the leader of the hamlet (*penghulu*) in order to gain permission to clear land. A study of Jambi shows the customary right to clear the forest and convert it to farmland is privately owned. People can claim rights to

Table 2.2 Area of land respondents owned in 2010–2014 according to acquisition type and year of acquisition (unit: ha) (Source of data: Authors' survey in 2010–2014)

Year of acquisition	Inherited	Purchase	Cleared	Distributed	Others	Total
Peatland	62.2	91	264.4	210.2	33.6	661.4
1965–1974	0.0	0.0	15.5	0.0	0.0	15.5
1975–1984	6.4	8.0	34.5	0.0	3.5	52.4
1985–1994	8.3	20.0	111.4	0.0	0.0	139.7
1995–2004	32.4	47.1	101	93.7	8.0	282.2
2005–2014	12.7	14.6	2.0	90.5	2.0	121.8
Unclassified	2.4	1.3	0.0	26.0	20.1	49.8
Non-peatland	60.2	45.9	63.1	2.5	10.6	182.3
1935–1944	1.2	0.0	0.0	0.0	0.0	1.2
1945–1954	2.0	0.0	0.0	0.0	0.0	2.0
1955–1964	0.0	1.0	0.0	0.0	0.0	1.0
1965–1974	0.0	0.0	10.5	0.0	0.0	10.5
1975–1984	22.4	11.0	17.5	0.0	3.5	54.4
1985–1994	9.3	9.3	21.8	0.0	0.0	40.4
1995–2004	12.2	18.3	13.3	0.5	0.5	44.8
2005–2014	9.1	6.0	0.0	2.0	0.2	17.3
Unclassified	4.0	0.3	0.0	0.0	6.4	10.7
Total	122.4	136.9	327.5	212.7	44.2	843.7

the land by building a sign using the branch of a tree. Swidden lands that go fallow after cultivation of dry rice revert to communal forest (Parlindungan 1978, pp. 15–16).

Table 2.2 shows the area of land according to how it was acquired and the year it was acquired.

Table 2.2 shows data only for land owned by present-day respondents. Therefore, if the parents of the respondent cleared the land in the 1930s and their daughter inherited the land in 1952—and still owns the land today—only the acquisition in 1952 is listed as inheritance in this table. The respondents made no mention of communal lands.[11]

---

[11] Masuda (2012) extensively discussed communal land (*tanah ulayat*) at a village in Pelalawan district, Riau province, Indonesia, based on fieldwork in 2000–2001 and 2003–2005. Studies on customary law have discussed the communal land in Jambi, Riau, and East Sumatra. For example, a collection of customary laws in Jambi (Adatrechtbundel 1912, pp. 199–205) showed that there were many types of uncultivated lands where customary community's disposal rights (*beschkkingsrecht*) were exercised, and all lands except the land where individual use rights appeared because of the clearing were thought as the land communal use rights were excised by the village (*doesoen*) or the district. Villagers who want to clear these lands should get permit from the head of the village or district. In the village surveyed for this paper, the village head mentioned that the land was communally possessed by the village. Much land ownership and stewardship is based on customary rights; however, no one claimed individual rights on the communal land in the household survey we conducted. Respondents only mentioned privately owned land and its utilization. This paper therefore discusses these privately owned lands.

Table 2.2 shows that present-day households inherited land obtained before national independence in 1945. It also shows that peatlands were already cleared during 1965–1974, and that the percentage of cleared land compared to land acquired by other means is decreasing over time. The percentage of non-peatland is higher for inherited lands than for others. For inherited, purchased, and cleared lands, many non-peatlands are found, yet almost all distributed lands, which were acquired after the advance of acacia timber plantations into the former peat swamp forest beginning in 1995, are peatlands. Inherited, purchased, and cleared land have basis in customary titling practices, so at least 63.4% of peatland and 92.7% of non-peatland were acquired on the basis of customary rights. As described above, most of the distribution took place after the intrusion of acacia plantations and illegal logging in the dried peat forest. People reported that some lands were also distributed by local government districts (*kabupaten*) promoting the planting of oil palm.

On land titling, the Prona program was implemented in the surveyed village, so many villagers have the associated land ownership certificates. Still, only some lands have been covered by the program. Some villagers have registered lands independently of the Prona program, and have corresponding certificates. In the field, many people possess land certification documents issued by the village office. This Letter of Land Statement (SKT) is issued by the village head. Sometimes the SKT is called *Surat Segel* because the seal of the government indicating payment of tax is printed on the letter. This letter originated from the colonial government policy in Sumatra as described earlier.

On land titling, in the field we find the Letter of Land Compensation (*Surat Ketrangan Ganti Kerugian* or SKGK) made at the time of transaction and signed by the land seller, the land buyer, the village head, head of the sub-village, head of the RW,[12] head of the RT, and owners of neighboring land. This letter of land compensation functions as an official document of land purchase in the field, just like the *Akta Jual Beli Tanah* formally issued by the PPAT.

As explained above, the head of sub-district, or *camat*, traditionally functions as a PPAT in rural areas. Some people therefore believe that the *camat* has more authority to certify land rights, so they have both SKT and SKGK signed by the *camat*.

The authors found four mains kinds of documents certifying land rights. The first is the Letter of Statement made by the village head (SKT) that has its origin in customary rights; the second is the SKT recognized by the head of the sub-district (SKT *camat*); the third is a letter related to the transaction, usually a purchase of land, the Letter of Land Compensation (SKGK). There are SKGKs signed by the *camat* and not signed by the *camat*. The fourth is a certificate of land ownership (*sertifikat tanah hak milik.*). The fourth document, a certificate of land ownership, clearly has the strongest authority to certify the land rights. It is based on the Torrens system that uses a cadastral map and is registered at the government agrarian office (*Badan Pertanahan Nasional*, BPN). The other letters are based on customary rights

---

[12] On RW and RT, see footnote 1.

Table 2.3 Land owned by respondents in 2010 and 2012 according to document certifying the land rights, year of acquisition, and peatland status (unit: ha) (Source of data: Authors' survey, 2010–2014)

Year of acquisition	No letter	SKT	SKT *camat*	SKGK[a]	Certificate	Others	Total
Peatland							
1965–1974	5.5	9.0	0.0	1.0	0.0	0.0	15.5
1975–1984	10.0	35.9	0.0	0.0	6.5	0.0	52.4
1985–1994	49.8	28.8	21.0	1.5	38.6	0.0	139.7
1995–2004	142.7	101.0	0.0	22.9	11.6	4.0	282.2
2005–2014	57.5	49.6	2.0	2.5	4.1	6.0	121.7
Unclassified	26.0	7.4	0.0	0.0	1.3	14.8	49.5
Sub-total	291.5	231.7	23.0	27.9	62.1	24.8	661.0
Non-peatland							
1935–1944	1.2	0.0	0.0	0.0	0.0	0.0	1.2
1945–1954	2.0	0.0	0.0	0.0	0.0	0.0	2.0
1955–1964	0.0	1.0	0.0	0.0	0.0	0.0	1.0
1965–1974	7.5	3.0	0.0	0.0	0.0	0.0	10.5
1975–1984	19.5	23.4	5.0	0.0	6.5	0.0	54.4
1985–1994	6.5	14.0	7.1	0.3	7.3	5.3	40.4
1995–2004	6.2	7.6	9.1	21.2	0.7	0.0	44.7
2005–2014	1.0	15.5	0.0	0.0	0.6	0.2	17.3
Sub-total	43.9	64.5	21.2	21.5	15.1	5.5	171.5
Total	335.4	296.2	44.2	49.4	77.2	30.3	832.5

[a]SKGK here includes those signed by the *camat* and those not signed by the *camat*, but signed by the village head

recognized by the people concerned, the village head, the head of the sub-village, the owner, and the people who own neighboring land.

The first three kinds of documents do not rely on cadastral survey maps but still have some authority, especially among villagers, and even in the government sector because these letters have been incorporated into the government administration as mentioned above. These letters are also easily contested especially by the government, especially in cases at the state forest.[13] Besides those letters, many villagers have no letters to certify their land rights. Once the rights of land are disputed, such people have no further documents demonstrating their land rights or claims.

Table 2.3 shows the area of land according to the type of document certifying the land rights, whether the land is peatland or not, and the year when the land was acquired.

Table 2.3 shows that among the areas reported by survey respondents, lands without any certifying document are the largest category. The amount of land

---

[13]Letter of Circulation No. 9/SE/V6/2013 issued by the head of national land body in 2013 (Surat Edaran No. 9/SE/VI/2013 Kepala Badan Tanah Nasional RI) defined the format of the letter of statement issued by the village head (SKT). The circulation said the land certified should be out of the government designated state forest area (*kawasan hutan*).

Table 2.4 Land owned by respondents in 2010–2011 according to peatland status, land use, and type of document certifying the land rights (unit: ha) (Source of data: Authors' survey, 2010–2014)

Land use	No letter	SKT	SKT camat	SKGK[a]	Certificate	Unclassified	Total
Peatland	291.5	231.6	23.0	27.9	62.2	10.5	646.7
Home garden	27.3	51.9	4.0	8.9	41.3	2.0	135.4
Farmland	264.2	179.7	19.0	19.0	20.9	8.5	511.3
Non-peatland	43.9	68.4	21.2	21.5	16.1	5.5	176.6
Home garden	30.0	25.5	14.1	3.5	13.6	0.3	87.0
Farmland	13.9	42.9	7.1	18.0	2.5	5.2	89.6
Total	335.4	300.0	44.2	49.4	78.3	16.0	823.3

[a]SKGK here includes those signed by the *camat* and those not signed by the *camat*

certified by the SKT is not small, however, while the lands recognized by government-issued land certificates are less than 10% of the total. Both peatlands and non-peatlands are recognized by these documents. The percentage of peatland is higher among lands lacking documentation and among lands with SKT. Lands without documents increased especially after 1995 following establishment of the acacia plantation in this area, but undocumented land was also found in earlier days (1935–1954). SKTs are found regardless of the year, and is especially indicative of land sales and acquisitions. The reason why so many lands are without letters will be discussed later.

Older people who owned land in 1935–1954 did not have any titling letters, but the rise in migration into the village and the subsequent increase in population and demand for land, seems to have incentivized interest in more secure land rights through documentation.

Table 2.4 shows land use (home garden [*pekarangan*] or farmland [*kebun*]), the type of document certifying land rights, and whether the land is peatland or non-peatland.

Table 2.4 shows that the percentage of non-peatland is higher among home gardens compared with farmlands. The percentage of land covered by land certificates is higher among home gardens compared with farmlands. All four types of documents are found for both peatland home gardens and non-peatlands home garden, while some people had no documentation at all.

2.3.2 Peatland Degradation and the Land Tenure System

Here we discuss peatland degradation in relation to the above-mentioned land conditions, such as peatland status, the type and year of land acquisition, and land titling. We first discuss the issue of burning, and second, the issue of land abandonment.

2.3.2.1 Peatland Burning on Land Owned by Respondents

As explained in Mizuno et al. (2016), peatland fire is a serious issue. How does this fire relate to the varieties of land mentioned above, such as the way the land is acquired and the status of the land title? (Tables 2.5 and 2.6)

We find some important trends in this Table 2.5. The percentage of peatlands burned is higher than that of non-peatlands. The percentage of burned lands is higher among those whose lands were acquired through distribution and clearing while the percentage is somewhat lower for inherited lands and purchased non-peatland. Even in non-peatland areas, however, a majority of the total area has burned. Here the data show how serious and extensive the issue of burning is.

In the following table we check the issue of burning according to the type of land certification letters (Table 2.6).

Table 2.5 Land owned by respondents in 2010 and 2011 according to type of acquisition, peatland status, and burned status (unit: ha) (Source of data: Authors' survey, 2010–2014)

Burned status	Inherited	Purchased	Cleared	Distributed	Unclassified	Total
Peatland	62.1	90.9	264.4	210.2	14.0	641.6
Not burned	8.3	14.2	4.0	22.0	4.5	53.0
Burned	53.8	76.7	260.4	188.2	9.5	588.6
Non-peatland	60.2	45.8	63.1	2.5	5.0	176.6
Not burned	30.3	26.8	15.8	0.5	1.5	74.9
Burned	29.9	19.0	47.3	2.0	3.5	101.7
Total	122.3	136.7	327.5	212.7	19.0	818.2
Not burned	38.6	41.0	19.8	22.5	6.0	127.9
Burned	83.7	95.7	307.7	190.2	13.0	690.3

Table 2.6 Land owned by respondents in 2010 and 2011 according to the document certifying the land rights, burned status, and peatland status (unit: ha) (Source of data: Authors' survey, 2010–2014)

Burned status	No letter	SKT	SKT Camat	SKGK[a]	Certificate	Unclassified	Total
Peatland	291.5	231.6	23.0	27.9	62.1	10.5	646.6
Not burned	6.0	32.4	3.0	2.9	4.2	4.5	53.0
Burned	285.5	194.2	20.0	25.0	57.9	6.0	588.6
Unclassified	0.0	5.0	0.0	0.0	0.0	0.0	5.0
Non-peatland	43.9	68.4	21.2	21.5	16.1	5.5	176.6
Not burned	18.2	23.4	2.1	16.8	8.9	5.5	74.9
Burned	25.7	45.0	19.1	4.7	7.2	0.0	101.7
Total	335.4	300.0	44.2	49.4	78.2	16.0	823.2
Not burned	24.2	55.8	5.1	19.7	13.1	10.0	127.9
Burned	311.2	239.2	39.1	29.7	65.1	6.0	690.3
Unclassified	0.0	5.0	0.0	0.0	0.0	0.0	5.0

[a]SKGK here includes those signed by the *camat* and those not signed by the *camat*

From Table 2.6, we can see that the percentage of land not burned is higher for non-peatlands with letters of certification and SKGK. In other words, certified lands with a letter of certificate and SKGK are burned somewhat less for non-peat land than those without. Despite this finding, the fact remains that burned lands are found among lands with all types of certifying documents, and among both peatlands and non-peatlands that do not have documents.

2.3.2.2 Land Abandonment in Relation to Land Use, Acquisition, Type, and Land Title

One of the causes of serious peatland degradation is land abandonment. A desolate landscape comprises extensive dried and barren peatland that is not cared for by the local people. This abandonment of peatland is both the reason for and result of peatland degradation and fire. The extensive area of abandoned land shown here is in accordance with the data on idle lands shown by Gaveau et al. (2016).

To understand the cause of land abandonment, we investigated the correlation between land use and land acquisition. Our hypothesis was that lands inherited by respondents from parents, or lands purchased by the respondents, would be better managed than land distributed by informal leaders. We supposed that people considered inherited or purchased lands as more socially and economically valuable, and that land certified with letters that strongly secured land rights would be better managed. On the other hand, we supposed that land with relatively weak titling would be less well managed.

Table 2.7 shows the relationship between land use land acquisition. Land use here is subdivided into land owned and operated, leased, fallow or reserved, and abandoned. In the case of fallow or reserved lands, people said that "*tanah itu sedang istirahat* [the land is taking a rest]" or "*belum pakai tanah itu* [the land is not utilized yet]." In the case of abandoned lands, people answered "*karena sering kebakaran, tanah itu jadi kosong* [because quite often the land is burnt, so now the land is empty]" or "*tanah itu kebakar terus, sekarang tidak tanam lagi* [the land is continuously burnt so now we do not plant on the land]," and so on.

According to Table 2.7, peatlands are abandoned far more frequently than non-peatlands. Lands acquired through distribution constitute the largest percentage of abandoned lands; cleared lands have the second highest rate of abandonment. On the other hand, lands acquired by inheritance or purchased were less commonly abandoned.

Now we turn to how land use relates to the type of document that certifies the land title (Table 2.8).

Among abandoned lands, 81.5% do have no corresponding certifying documents, while almost no land with a SKT *camat*, SKGK, or a certificate is abandoned, according to Table 2.8. On the other hand, owned and operated lands have various certifying documents, and some have no documents.

How about the relationship among the burning experience, land use, and documents to certify the land rights? Although peatland burning is a serious issue in the

Table 2.7 Land owned by respondents in 2010 and 2011 according to land use and when it was acquired (unit: ha) (Source of data: Authors' survey, 2010–2014)

Acquisition type	Owned and operated	Leased	Fallow or reserved	Abandoned	Unclassified	Total
Peatland	369.4	3.5	17.0	255.7	1.0	646.6
Inherited	55.1	0.0	0.0	6.0	1.0	62.1
Purchased	75.7	0.0	2.0	13.2	0.0	90.9
Cleared	153.4	0.0	0.0	111.0	0.0	264.4
Distributed	80.2	0.0	8.5	121.5	0.0	210.2
Unclassified	5.0	3.5	6.5	4.0	0.0	19.0
Non-peatland	145.8	5.8	16.0	9.0	0.0	176.6
Inherited	50.5	0.0	8.6	1.0	0.0	60.1
Purchased	33.4	0.0	5.4	7.0	0.0	45.8
Cleared	60.9	2.3	0.0	0.0	0.0	63.2
Distributed	0.5	0.0	2.0	0.0	0.0	2.5
Unclassified	0.5	3.5	0.0	1.0	0.0	5.0
Total	515.2	9.3	33.0	264.7	1.0	823.2

Table 2.8 Land owned by respondents in 2010 and 2011 according to land use and documents certifying the land title (unit: ha) (Source of data: Authors' survey, 2010–2014)

Type of certifying document	Owned and operated	Leased	Fallow or reserved	Abandoned	Unclassified	Total
Peatland	369.4	3.5	17.0	255.7	1.0	646.6
No letter	74.0	0.0	2.0	215.5	0.0	291.5
SKT	186.4	3.5	6.5	34.2	1.0	231.6
SKT *camat*	21.0	0.0	2.0	0.0	0.0	23.0
SKGK[a]	23.9	0.0	0.0	4.0	0.0	27.9
Certificate	60.1	0.0	2.0	0.0	0.0	62.1
Unclassified	4.0	0.0	4.5	2.0	0.0	10.5
Non-peatland	145.9	5.8	16.0	9.0	0.0	176.7
No letter	36.8	0.0	6.1	1.0	0.0	43.9
SKT	58.8	5.5	2.1	2.0	0.0	68.4
SKT *camat*	21.2	0.0	0	0.0	0.0	21.2
SKGK[a]	16.2	0.0	0.3	5.0	0.0	21.5
Certificate	12.6	0.0	2.5	1.0	0.0	16.1
Unclassified	0.3	0.3	5.0	0.0	0.0	5.6
Total	515.3	9.3	33.0	264.7	1.0	823.3

[a]SKGK here includes those signed by the *camat* and those not signed by the *camat*

Table 2.9 Burned land owned by respondents in 2010 and 2011 according to land use and land title documents (unit: ha) (Source of data: Authors' survey 2010–2014)

	Owned and operated	Leased	Fallow or reserved	Abandoned	Unclassified	Total
No letter	92.1	0.0	3.6	215.5	0.0	311.2
SKT	193.0	7.0	2.0	36.2	1.0	239.2
SKT *camat*	37.1	0.0	2.0	0.0	0.0	39.1
SKGK[a]	25.7	0.0	0.0	4.0	0.0	29.7
Certificate	63.1	0.0	2.0	0.0	0.0	65.1
Unclassified	4.0	0.0	0.0	2.0	0.0	6.0
Total	415.0	7.0	9.6	257.7	1.0	690.3

[a]SKGK here includes those signed by the *camat* and those not signed by the *camat*

study village as has been shown at Table 2.3, not all lands have been burned. So how have the people responded to the burning of land?

Table 2.9 shows the relationship between land use and land titles for burned land.

As has demonstrated in Table 2.3, burning is extensive and a serious problem in the research site, leading to the abandonment of several lands. As Table 2.9 shows, of the entire burned area of 690.2 ha, as much as 257.7 ha (37.3%) have been abandoned, while 414.9 ha (60.1%) are still owned and operated. Many factors would influence these outcomes, as will be comprehensively analyzed in the conclusion. Here we examine only the relationships between burning and kind of documents certifying land rights. Of abandoned burned land, 83.6% (215.5 ha) has no certifying letter, while 36.2 ha (14.0%) have SKT. Of burned land with titling certificates, none has been abandoned. Burned land that is still managed (414.9 ha in total) has many variations of land titling. Some of this land has no corresponding letter of certification, but a majority of burned but operated land does have some kinds of letter to certify the land rights.

Figure 2.1 shows the relationship between land abandonment and letters to certify the land rights.

As discussed above, the government-issued land certificate is the strongest kind of certification, while having no letter at all is the weakest. Government land certificates are based on cadaster survey and so is integrated into the land registration of National Agrarian office. On the other hand, SKT is not based on cadaster survey, so it is said to bear less authority than the land certificate as land title. Although SKT is not based on the cadaster survey, the latter was recognized by the head of village office and in many cases by the head of sub-district office. The SKT made by the head of sub-district (the SKT authorized by head of sub-district was authorized by the head of village beforehand) carries greater authority than the SKT made only by the village head. SKGK is the letter to certify the transaction of land assuming the land certificate would be made, so the SKGK is thought to be a letter under the system of land certificate, even though it is not based on cadaster survey.

As a consequence, we can assume that following the land certificate, the second most secure is the SKGK, the SKT authorized by the head of sub-district, and the SKT authorized by the village head come in third. The weakest is no letter. From

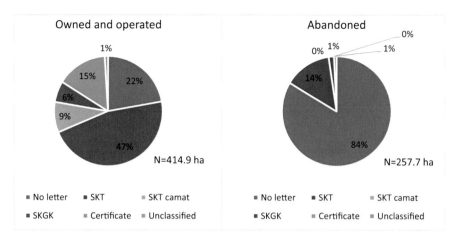

Fig. 2.1 Relationship between land titling and peatland abandonment at surveyed village

Fig. 2.1, we can say that under the strongest land right, more farmers continued to manage their lands even after fire. The weaker the land rights, the more likely the land is abandoned after burning.

Two categories of land, abandoned peatland and managed land, both exist entirely without any certifying documents. In order to better understand meaning of these categories, we analyze the year the lands were acquired in Table 2.10.

Table 2.10 shows that almost all abandoned lands have been acquired relatively recently, especially since 1995, and that almost no abandoned lands were acquired prior to 1984. On the other hand, lands that are owned and operated today have often been acquired since village establishment.

Table 2.3 shows that many actively managed lands without title were acquired from the 1930s to 1984. This is because traditional Malays often felt no need to seek any land title, and have remained so on their lands since the early days without title of any kind.

2.3.3 Analysis and Discussion of Abandoned Land

Abandonment of land is one of the most apparent manifestations of peatland degradation and has led to increasing incidence of fire, as explained by Joosten et al. (2012). To curb further land abandonment, we should understand why lands are abandoned in the first place.

We have seen how land titling influences abandonment of burned land. This section will present a more comprehensive discussion. So far we understand that many factors are related to land abandonment besides land title. Land cover type (whether the land is peatland or not), burn status, acquisition process, and year of acquisition all seem to be related land abandonment. Although there are many plots

Table 2.10 Land owned by the respondents in 2010 and 2011 according to acquisition type, peatland status, and year of land acquisition (unit: ha) (Source of data: Authors' survey, 2010–2014)

Year of land acquired	Owned and operated	Leased	Fallow or reserved	Abandoned	Unclassified	Total
Peatland	369.5	3.5	17.0	255.7	1.0	646.7
1965–1974	15.5	0.0	0.0	0.0	0.0	15.5
1975–1984	47.9	3.5	0.0	1.0	0.0	52.4
1985–1994	95.7	0.0	0.0	44	0.0	139.7
1995–2004	158.0	0.0	6.0	118.2	0.0	282.2
2005–2011	43.8	0.0	10.5	66.5	1.0	121.8
Unclassified	8.6	0.0	0.5	26.0	0.0	35.1
Non-peatland	145.9	5.8	16.0	9.0	0.0	176.7
1935–1944	1.2	0.0	0.0	0.0	0.0	1.2
1945–1954	2.0	0.0	0.0	0.0	0.0	2.0
1955–1964	1.0	0.0	0.0	0.0	0.0	1.0
1965–1974	10.5	0.0	0.0	0.0	0.0	10.5
1975–1984	46.4	3.5	4.5	0.0	0.0	54.4
1985–1994	28.4	2.3	7.8	2.0	0.0	40.5
1995–2004	38.0	0.0	1.7	5.0	0.0	44.7
2005–2011	14.1	0.0	2.0	1.0	0.0	17.1
Unclassified	4.3	0.0	0.0	1.0	0.0	5.3
Total	515.4	9.3	33.0	264.7	1.0	823.4

of abandoned land, there are as many plots that have not been abandoned although the lands experienced burning. What factors determine whether the plot is abandoned or not? Here we use multiple co-regression to analyze the factors that determine abandonment. Whether a plot is abandoned or not is qualitative data that can be analyzed with a probit model.

Among the variables showing as important factors above, we review the following: peatland status, experience of fire, type of acquisition (especially distribution), documents certifying land rights, year of acquisition, and the area of land plots. Peatland status is judged by the respondents. If a plot comprised part peatland and part non-peatland, we divided the plot and counted it as two plots. We asked the respondent the history of fire for each plot. If a land has been burnt, it was classified as "burned." Table 2.7 shows that land acquired through distribution most often correlates to land abandonment. We therefore classified land acquisition into two types only, that is, by distribution or not. Whether lands were with certifying documents, or not is important factor that influence on land abandonment as discussed earlier, here we employing the factor of certifying document by classifying into two categories: those with documents and those without. This was done because as long as a person has a document of certification, they can claim their right to the land if there is a dispute, but if there is no document at all it is difficult to claim rights to the land. All documents also have legal basis in either customary law or public law as has been shown. The availability of certifying documents represents the factor of ownership, which is important because in cases of leasing land, such as sharecropping. Lands under those arrangements were not included because share-croppers do not know about, or are not often familiar with, official documentation or history of fire. According to Table 2.10, year of land acquisition seems also to be closely related to whether the land is abandoned or not. The size of the land plot may be influential as well because if the land is too large, it may not be easily managed.

Some correlations among these factors seem to exist. We analyzed the correlation between the variables of the year of acquisition, burned status, land area, and the document certifying the land right.[14] Peatland status is correlated with fire history and acquisition type is correlated with the year of acquisition. We therefore dropped the variables of peatland status and acquisition type to avoid the problem of multicollinearity. The variable of fire history therefore represents both the factors of peatland (or not) and of fire (burned or not), and the variable of land acquisition year represents both the way the land was acquired and the year it was acquired.

Considering the factors mentioned above, we have assumed the following linear equation:

$$DA = F1\ (YEA, FIR, ARE, DOC)$$

In order to examine the influence of these factors on land abandonment, we have assumed the following linear equation parameters:

DA: Dummy variable of abandonment of land for each plot (1: Abandoned in, 0: Abandoned out)

YEA: Year of land acquisition

FIR: Dummy variable for the burned status of each plot (1: Experience of burning in, 0: Experience of burning out)

ARE: Area of land of each plot (in hectares)

---

[14]Correlation among the variables is as follows:

	Peatland status	Year of acquisition	Burned status	Area of plot	Type of acquisition	Document certifying the land right
Peatland status	1	0.2968	0.3682	0.1577	0.3710	−0.0765
The year of acquisition	0.2968	1	0.0426	− 0.0138	0.4462	−0.0271
Burned status	0.3682	0.0426	1	0.1843	0.2186	−0.1155
Area of plot	0.1577	−0.0138	0.1843	1	0.0071	−0.1345
Type of acquisition	0.3710	0.4462	0.2185	0.0071	1	−0.3754
Document certifying the land right	−0.0765	−0.0271	− 0.1155	− 0.1345	−0.3754	1

Table 2.11 Results of probit model estimation

Variable	Coefficient	Std. Error	z-Statistic	Prob.
C	−133.7682[a]	28.6678	−4.6662	0.0000
Year of land acquisition	0.0674[a]	0.0143	4.6409	0.0000
Burned status	1.1345[a]	0.3252	3.4885	0.0005
Area of land	0.0131	0.0199	0.6573	0.5110
Document certifying the land rights	−1.2178[a]	0.2303	−5.2881	0.0000
McFadden R-squared	0.365799			
Log likelihood	−78.42526			
Total obs	238			
Obs with Dep = 0	187			
Obs with Dep = 1	51			

[a]= 1% significance level

DOC: Dummy variable for document certifying the land title (1: With any kind of document in, 0: Without any document out).

As Table 2.11 shows, the variables of year of land acquisition, burned status, and document certifying the land rights are significantly correlated to land abandonment (all variables are significant with level of 1%). On the other hand, the area of land has no correlation with abandonment. These factors explain as much as 37% of abandonment. This means that whether land is peatland or not, manner of land acquisition (especially through distribution), and the existence of documents certifying land rights are significantly correlated to land abandonment.

Although fire is quite common in the surveyed village, the year of land acquisition, the type of land acquisition, and the existence of documents to certify land rights is much more determinant of whether land is abandoned. Lands distributed after 1985, and especially after 1995, were often peatlands and were frequently abandoned after fire. However, lands acquired prior to 1984, and acquired by inheritance, purchasing and clearance have typically not been abandoned, although many of them are also peatlands and have experienced fire.

2.4   Discussion and Conclusion

Gaveau et al. (2016) have shown that though much plantation land has burned, far more land outside of concession areas has burned. What is the relationship among the timber plantation, drainage, peatland degradation, in-migration, fire and abandonment of peatland?

At the research site, our surveys show that serious peatland degradation and fire began to take place since the end of the 1990s. Since the middle of the 1990s timber plantations started operations requiring large-scale drainage that dried up land outside the concession, leaving it vulnerable to fire.

Observing fire in the peat swamp forest outside the concession, local people began cutting trees in the forest and sold timber to Malaysia when the rule of law loosened after President Suharto stepped down. Once the timber companies began to drain peatlands, so beginning the occurance of fire, people moved into the forest area to cut trees and distributed the land among themselves. They did this partly for profit and partly to secure land rights. People thought that if they did not secure land rights in the peat swamp forest, even more land would be given to the companies as concession.

In this situation, securing land rights promoted peatland degradation because after securing the peatland many of these small land holders planted oil palm, often only to abandon the land later. This information on the history of peatland degradation obtained from local people accorded with the finding from the multiple correlation analysis—that there is a greater likelihood for land to be abandoned when it has been acquired through distribution.

In-migration to the area began in 1930, but no associated peatland degradation was found. On the other hand, when the economic activities were enhanced especially since the middle of the 1990s, many non-Malay couples immigrated to the village, also contributing to peatland degradation.

Clearing, inheritance, and purchase of land began at the beginning of the 1930s, while land distribution began typically in the middle of the 1990s. More people thus began to secure land rights by getting land titles. These changes, especially from the middle of the 1990s, have encouraged the Malay people to cultivate the peatland intensively by planting oil palm or rubber.

The results of this study clearly show that people who have secured land rights by obtaining land titles tend to better manage the peatland, while those without land titles tend to abandon the land after burning. Thus securing land rights on peatland has two different consequences. First, the motivation to secure land rights has promoted further land distribution when there is competition in the peat swamp forest between people and companies. Second, once people have acquired peatland, the stronger the land right, the better the peatland management.

The foregoing discussion should also be related to the state forest. One government document states that many SKT documents have been issued for the Government-designated state forest area, or state forest area.[15] As per the current government forestry policies, SKT and SKGK issued on state forest area tend not to be recognized. The village heads and *camat* are prohibited from issuing the SKT and SKGR.[16] But the boundary of the state forest area is not clear. When interviewed, the village head said that he did not know the boundary.[17] The former vice village head

---

[15] Badan pertanahan Nasional, Petunjuk Pelaksanaan Kegiatan Inventarisari Penguasaan, Pemilikan, Penggunaan dan Pemanfaatan Tanah (IP4T) Dalam Kawasan Hutan Jakarta: Kementrian Agrarian dan Tata Ruang, Badan Pertanahan Nasional, 2015.

[16] See footnote 11. Tribune Pekanbaru, March, 27, 2013 'KADES dan CAMAT Jangan kelaurkan SKT dan SKGR (Village officers and *camat* are prohibited from issuing the SKT and SKGR, http:// pekanbaru.tribunnews.com/2013/03/27/camat-dan-kades-jangan-keluarkan-skt-dan-skgr).

[17] Authors' interview with village head on 28 March 2011, in the surveyed hamlet.

said the land 1 km inland of the seashore is government-designated state forest.[18] Another former village officer reported that the area within a radius of 5 km from the seashore is people's land, and land further inland of that is government-designated state forest.[19] Data shown above on the area of the state forest (Mizuno et al. 2021) is only the data based on the map shown on the BRGM home page in 2020, but it was not shared with local people, including the village head.

Since around 2000, distributed land was located further inland, so the village office was thought to be reluctant to issue the SKT. This ambiguity of state land borders is also one reason why so many distributed lands did not have proper documentation. Logging in the peat swamp forest outside the concession and distribution of land in the 2000s took place in the state forest. Weak state management of its forest areas and the ambiguity of boundaries facilitated the intrusion of timber plantations and, consequently, the degradation of peatland and abandonment of burned peatland. This condition of the state forest is not unique to the research site because state forest covers 65% of the Indonesian land area, and all timber plantations operate in state forest areas.

A confluence of factors contribute to peatlands degradation. This study shows that past degradation can be associated with specific terms of land use and title. Importantly, this study also shows that good peatlands management has been supported by secure land titling, especially those that support longstanding customary land practices, and can do so again.

References

Adatrechtbundel (1912) Adatrechtbundel V, Bezorgd door de commissie voor het adatrecht (Gemengd), Het Koninklijk Instituut voor de Taal, Land, en Volkskunde van Nederlandsch-Indie. Martinus Nijhoff, 's-Gravenhage

Adatrechtbundel (1916) Adatrechtbundels XII: Bezorgd door de Commissie voor het Adatrecht en uitgegeven door het Het Koninklijk Instituut voor de Taal, Land, en Volkskunde van Nederlandsch-Indie. Gemengd, Martinus Nijhoff, 's-Gravenhage

Arisaputra MI (2015) Reforma agraria di Indonesia. Sinar Grafika, Jakarta

Bukit Batu sub-district (2010) Monograf Bukut Batu sub-district, Bukit-batu sub-district (Kecamatan Bukut Batu). Pemerintah Kecamatan Bukit Batu, Kabupaten Bengkalis, Propinsi Riau

Carlson KM, Curran LM, Asner GP et al (2013) Carbon emissions from forest conversion by Kalimantan oil palm plantations. Nat Clim Chang 3(3):283–287. https://doi.org/10.1038/nclimate1702

Chokkalingam U, Kurniawan I, Ruchiat Y (2005) Fire, livelihoods, and environmental change in the Middle Mahakam peatlands, East Kalimantan. Ecol Soc 10(1):26. https://doi.org/10.5751/ES-01320-100126

---

[18] Authors' interview with respondent No. 60 on 24 December 2014, in the surveyed hamlet.

[19] Authors' interview with respondent No. 65 on 30 March 2011, in the surveyed hamlet.

Departement van Landbouw, Nijverheid, en Handel (1926) De bevolkingsrubbercultuur in Nederlandsch-Indië: VI. riouw en onderhoorigheden, oostkust van Sumatra en Atjeh en onderhoorigheden. Landsdrukkerij, Weltevreden

Dohong A, Aziz AA, Dargusch P (2017) A review of the drivers of tropical peatland degradation in South-East Asia. Land Use Policy 69:349–360. https://doi.org/10.1016/j.landusepol.2017.09.035

Endert FH (1932) De proefbaanmetingen in de panglong-gebied van Bengkalis (Sumatra's Oostkust) en Riouw. Tectona XXV:731–787

Feder G, Feeny D (1993) The theory of land tenure and property rights. In: Hoff K, Braverman A, Stiglitz JE (eds) The economics of rural organization: theory, practice, and policy. Oxford University Press, Oxford

Feder G, Noronha R (1987) Land rights systems and agricultural development in Sub-Saharan Africa. World Bank Res Obs 2(2):143–169

Fuller DO, Hardiono M, Meijaard E (2011) Deforestation projections for carbon-rich peat swamp forests of Central Kalimantan, Indonesia. Environ Manag 48(3):436–447. https://doi.org/10.1007/s00267-011-9643-2

Furukawa H (1992) Indoneshia no teishicchi. Keiso Shobo, Tokyo. [English edition: Furukawa H (1994) Coastal wetlands of Indonesia: environment, subsistence and exploitation (trans: Hawkes PW). Kyoto University Press, Kyoto]

Galudra G, van Noordwijk M, Suyanto S et al (2011) Hot spots of confusion: contested policies and competing carbon claims in the peatlands of Central Kalimantan, Indonesia. Int For Rev 13(4):431–441. https://doi.org/10.1505/146554811798811380

Gaveau DLA, Pirard R, Salim MA et al (2016) Overlapping land claims limit the use of satellites to monitor *no-deforestation* commitments and *no-burning* compliance. Conserv Lett 10(2):257–264. https://doi.org/10.1111/conl.12256

Harahap A (1991/1992) Hak atas tanah di dalam konsolidasi. Pusat Penelitian dan Pengembangan Badan Pertanahan Nasional, Jakarta

Hooijer A, Silvius M, Wösten H et al (2006) PEAT-CO$_2$: assessment of CO$_2$ emissions from drained peatlands in SE Asia. Delft Hydraulics Report Q3943, Deltares, the Netherlands

Jelles JGG (1929) De boschexploitatie in het panglonggebied. Tectona XXII:482–505

Jewitt SL, Nasir D, Page SE et al (2014) Indonesia's contested domains: deforestation, rehabilitation and conservation-with-development in Central Kalimantan's tropical peatlands. Int For Rev 16(4):405–420. https://doi.org/10.1505/146554814813484086

Joosten H, Tapio-Biström ML, Tol S (2012) Peatlands: guidance for climate change mitigation through conservation, rehabilitation and sustainable use. In: Mitigation of climate change in agriculture series 5, 2nd edn. FAO/Wetlands International, Rome/Wageningen

Kunz Y, Steinebach S, Dittrich C et al (2017) 'The fridge in the forest': historical trajectories of land tenure regulations fostering landscape transformation in Jambi Province, Sumatra, Indonesia. For Policy Econ 81:1–9. https://doi.org/10.1016/j.forpol.2017.04.005

Law EA, Bryan BA, Meijaard E et al (2015) Ecosystem services from a degraded peatland of Central Kalimantan: implications for policy, planning, and management. Ecol Appl 25(1):70–87. https://doi.org/10.1890/13-2014.1

Maimunah S, Rahman SA, Samsudin YB et al (2018) Assessment of suitability of tree species for bioenergy production on burned and degraded Peatlands in Central Kalimantan, Indonesia. Land 7(4):115. https://doi.org/10.3390/land7040115

Masuda K (2012) Indoneshia mori no kurashi to kaihatsu: tochi wo meguru 'tsunagari' to 'semegiai' no shakaishi (Indonesia, livelihood at forest and development, social history of the linkage and negotiation relating to the land). Akashi shoten, Tokyo

Masuda K, Kusumaningtyas R, Mizuno K (2016) Local communities in the peatland region: demographic composition and land use. In: Mizuno K, Fujita MS, Kawai S (eds) Catastrophe and regeneration in Indonesia's peatlands: ecology, economy and society. Kyoto CSEAS series on Asian studies, vol 15. NUS Press/Kyoto University Press, Singapore/Kyoto, pp 185–210

Medrilzam M, Dargusch P, Herbohn J et al (2014) The socio-ecological drivers of forest degradation in part of the tropical peatlands of Central Kalimantan, Indonesia. Forestry 87(2):335–345. https://doi.org/10.1093/forestry/cpt033

Medrilzam M, Smith C, Aziz AA et al (2017) Smallholder farmers and the dynamics of degradation of peatland ecosystems in Central Kalimantan, Indonesia. Ecol Econ 136:101–113. https://doi.org/10.1016/j.ecolecon.2017.02.017

Miettinen J, Liew SC (2010) Degradation and development of peatlands in Peninsular Malaysia and in the islands of Sumatra and Borneo since 1990. Land Degrad Dev 21(3):285–296. https://doi.org/10.1002/ldr.976

Miettinen J, Hooijer A, Shi C et al (2012a) Extent of industrial plantations on Southeast Asian peatlands in 2010 with analysis of historical expansion and future projections. GCB Bioenergy 4(6):908–918. https://doi.org/10.1111/j.1757-1707.2012.01172.x

Miettinen J, Shi C, Liew SC (2012b) Two decades of destruction in Southeast Asia's peat swamp forests. Front Ecol Environ 10(3):124–128. https://doi.org/10.1890/100236

Mizuno K (1991) Nishi Jawa nouson niokeru tochi shoyuken no kakunin shorui hoyu jokyo (Right of land ownership and documents to certify its right in rural West Java). In: Umehara H (ed) Tonan-ajia no tochi seido to nogyo henka (Land tenure and agricultural change in Southeast Asia), IDE Research Series, no 406. Institute of Developing Economies, Tokyo, pp 251–308. https://doi.org/10.20561/00044457

Mizuno K (1996) Rural industrialization in Indonesia: a case study of community based-weaving industry in West Java, Occasional Paper Series no 31. Institute of Developing Economies, Tokyo

Mizuno K, Shigetomi S (eds) (1997) Tonan-ajia no keizai kaihatsu to tochi seido (Economic development and land system in South-East Asia), IDE Research Series, no 477. Institute of Developing Economies, Tokyo

Mizuno K, Fujita MS, Kawai S (2016) Catastrophe and regeneration in Indonesia's peatlands: ecology, economy and society, Kyoto CSEAS series on Asian studies, vol 15. NUS Press/Kyoto University Press, Singapore/Kyoto, p 466

Mizuno K, Hosobuchi M, Ratri DAR (2021) Land tenure on peatland: a source of insecurity and degradation in Riau, Sumatra. In: Osaki M, Tsuji N, Foead N et al (eds) Tropical peatland eco-management. Springer, Singapore, pp 627–649. https://doi.org/10.1007/978-981-33-4654-3_23

Momose K (2002) Environments and people of Sumatran peat swamp forests II: distribution of villages and interactions between people and forests. Southeast Asian Stud 40(1):87–108

Muchsin, Koeswahyuno I, Soimin (2019) Hukum agraria Indonesia dalam perspektif sejarah. Refika, Jakarta

Nawir AA, Murniati, Rumboko L (eds) (2007) Forest rehabilitation in Indonesia: where to after more than three decades? CIFOR, Bogor. https://doi.org/10.17528/cifor/002274

Page S, Hosciło A, Wösten H et al (2009) Restoration ecology of lowland tropical peatlands in Southeast Asia: current knowledge and future research directions. Ecosyst 12(6):888–905. https://doi.org/10.1007/s10021-008-9216-2

Parlindungan AP (1978) Pandangan kritis sebagai aspek dalam pelaksanaan Undang-undang pokok Agraria di daerah Jambi. Alumni, Bandung

Pastor G (1927) De Panlongs. Publicaties van het Kantoor van Arbeid, vol 3. Landsdrukkerij, Weltevreden

Resosudarmo IAP, Atmadja S, Ekaputri AD et al (2014) Does tenure security lead to REDD+ project effectiveness? Reflections from five emerging sites in Indonesia. World Dev 55:68–83. https://doi.org/10.1016/j.worlddev.2013.01.015

Resosudarmo IAP, Tacconi L, Sloan S et al (2019) Indonesia's land reform: Implications for local livelihoods and climate change. For Policy Econ 108:101903. https://doi.org/10.1016/j.forpol.2019.04.007

Rietberg PI, Hospes O (2018) Unpacking land acquisition at the oil palm frontier: obscuring customary rights and local authority in West Kalimantan, Indonesia. Asia Pac Viewp 59(3): 338–348. https://doi.org/10.1111/apv.12206

Sabiham S (2009) Pengolahan lahan gambut Indonesia berbasis keunikan ekosistem. Institut Pertanian Bogor Press, Bogor

SarVision (2011) Impact of oil palm plantations on peatland conversion in Sarawak 2005–2010. Summary report Version 1.6. SarVision, Wageningen

Schadee WHM (1918) Geschidenis van Sumatra's oostkust, Medeling no 2. Oost-kust van Sumatra-Instituut, Amsterdam

Sewandono M (1937) Inventarisatie en inrichting van de veenmoerasbosschen in het panglong-gebied van Sumatra's Oostkust. Tectona XXX:660–679

Sheil D, Casson A, Meijaard E et al (2009) The impacts and opportunities of oil palm in Southeast Asia: what do we know and what do we need to know? Occasional Paper no 51. CIFOR, Bogor. https://doi.org/10.17528/cifor/002792

Silvius MJ, Suryadiputra N (2005) Review of policies and practices in tropical peat swamp forest management in Indonesia. Wetlands International, Netherlands

Stone R (2007) Can palm oil plantations come clean? Science 317(5844):1491. https://doi.org/10.1126/science.317.5844.1491

Suyanto S, Premana RP, Khususiyah N et al (2005) Land tenure, agroforestry adoption, and reduction of fire hazard in a forest zone: a case study from Lampung, Sumatra, Indonesia. Agrofor Syst 65(1):1–11. https://doi.org/10.1007/s10457-004-1413-1

Venter O, Meijaard E, Wilson K (2008) Strategies and alliances needed to protect forest from palm-oil industry. Nature 451:16. https://doi.org/10.1038/451016a

Veth PT (1869) Aardrijkskundig en staatistisch woordenboek van Nederlandsch-Indië. Van Kampen, Amsterdam

Watanabe K, Masuda K, Kawai S (2016) Deforestation and the process of expansion of oil palm and acacia plantations. In: Mizuno K, Fujita MS, Kawai S (eds) Catastrophe and regeneration in Indonesia's peatlands: ecology, economy and society, Kyoto CSEAS series on Asian studies, vol 15. NUS Press/Kyoto University Press, Singapore/Kyoto, pp 281–295

Wildayana E, Armanto ME, Zahri I et al (2019) Socio economic factors causing rapid peatlands degradation in South Sumatra. Sri J Env 3(3):87–95. https://doi.org/10.22135/sje.2018.3.3.87-95

Yusran Y, Sahide MAK, Supratman S et al (2017) The empirical visibility of land use conflicts: from latent to manifest conflict through law enforcement in a national park in Indonesia. Land Use Policy 62:302–315. https://doi.org/10.1016/j.landusepol.2016.12.033

Chapter 3
Characteristics of Bird Community Response to Land Use Change in Tropical Peatland in Riau, Indonesia

Motoko S. Fujita, Hiromitsu Samejima, Dendy Sukma Haryadi, and Ahmad Muhammad

Abstract Tropical peatlands have increasingly been targeted for logging or conversion to plantations in recent years. Tropical peatlands are unique ecosystems rich in biodiversity, but they have not attracted as much researcher attention as tropical forests, for example. There is still limited understanding of the ecological significance of peatland disturbance, or of the ecological resilience of peatland ecosystems. This study focuses on birds as indicator species in peatland ecologies. It compares bird communities in peatlands and non-peat lowlands in terms of: (1) species richness; (2) feeding guilds; and (3) responses to disturbance. Our research team analyzed bird communities in peatlands under several different land uses in Riau in comparison to those living in non-peat lowlands in Sumatra Island. We found that species richness in natural forests was lower in peatlands than in non-peat lowlands. The Jackknife estimator of species richness was 77.2 in natural forests on peatland, whereas on non-peat lowland, it was 114.8 and 241. Compared to non-peat lowland forests, the number of terrestrial insectivore and woodpeckers was lower in peatlands. Non-metric multidimensional scaling (NMDS) analysis showed that the bird community composition in peatland forests is unique compared to non-peat lowland forests, as they showed completely different lines of avifauna. Nevertheless, avifauna in disturbed sites on peatland were close to avifauna in disturbed non-peat lowland sites, which indicates that the disturbance of peatland would lead to homogenization of avifauna and loss of uniqueness, which in turn, leads to loss of biodiversity. Bird community composition in peatlands was very sensitive to land use change. Shifts in the community composition along the disturbance, as measured

M. S. Fujita (✉)
Center for Southeast Asian Studies, Kyoto University, Kyoto, Japan

H. Samejima
Institute for Global Environmental Strategies, Hayama, Kanagawa, Japan

D. S. Haryadi
Bogor Agricultural University, Bogor, West Java, Indonesia

A. Muhammad
University of Riau, Pekanbaru, Riau, Indonesia

by Euclidean distances in the NMDS plot between each disturbed habitat and natural forest, were greater in peatland than in non-peat lowland. Although our knowledge and data of peatland ecologies are limited, it seems likely that certain peatland avifauna can only survive in natural peat swamp forest.

Keywords Biodiversity · Tropical forest · Peat swamp · NMDS ordination · Disturbance

3.1 Introduction

Birds are known to contribute to numerous ecosystem services, including pollination, pest control, and seed dispersal, which benefit both human welfare and economy at local and global scales (Millennium Ecosystem Assessment 2005; Şekercioğlu et al. 2016). Birds therefore serve as good indicators of overall ecosystem status (O'Connell et al. 2000). Sumatra Island has the largest and richest bird diversity of all the Indonesian Islands, with 580 known bird species, 21 of which are endemic species, and 120 migrant species (Whitten et al. 2000). Representative vegetation in the island consists mainly of lowland mixed dipterocarp forests, tropical montane forests, mangroves, and peat swamp forests. The original area of Sumatran peatland is estimated as 7.3–9.7 million ha, accounting for one-quarter of the world's tropical peat swamp forests (Whitten et al. 2000), creating an important refuge for the birds of Sumatra (Yule 2010).

Lowland dipterocarp forests on the island have decreased dramatically in the last several decades due to development (Whitten et al. 2000; Laumonier et al. 2010), while a large area of natural forest remained in montane regions or peatlands. More recently peatland has become the target of commercial logging and conversion to plantations, especially for production of oil palm and acacia (Uryu et al. 2008; Corlett 2009; Koh et al. 2011). Peatland forests have attracted several basic studies on tree diversity (Anderson 1961; Bruenig 1990; Posa et al. 2011) and animal diversity (Whitemore 1984; Gaither Jr 1994; Philips 1998; Cheyne et al. 2010; Cheyne and Macdonald 2011; Posa 2011), but the impact of land use change on peatland ecosystems has been largely overlooked.

Posa (2011) reported that avian species diversity in selectively logged peatland forests in Central Kalimantan was quite similar to that found in intact forests. Avian diversity decreases by half in highly logged forest or recently burned brushland, however. Forest conversion, which usually replaces native trees with commercial trees or crops, generally has an even more significant effect on biodiversity (Barlow et al. 2007). For example, in Riau, Sumatra, Fujita et al. (2016) reported that the conversion of peat swamp forest into planted acacia forest and planted rubber forest had significant effects on bird diversity.

How resilient are peatland forests to forest degradation and conversion? What is their capacity recover desirable ecosystem states along with associated ecosystem services? Biodiversity is said to enhance ecosystem resilience (Elmqvist et al. 2003). In comparison to non-peat lowland forest, peat swamp forest supports low animal density and diversity (Janzen 1974; Whitten et al. 2000; Posa et al. 2011) due to its low nutrient content and low plant productivity (Bruenig and Droste 1995). Gaither Jr (1994) reported that although bird populations in peat swamp forest had low species diversity and density overall, some species have a higher density than in lowland dipterocarp forests. The uniqueness of the peatland ecosystem is attributed to a swampy ecosystem that is formed on peat soil, which contains high levels of organic carbon from undecomposed deadwood. Peatland soils are usually water-logged and therefore have high acidity (pH 3–4.5) and low nutrient content.

Ecosystem or community resilience is evaluated by comparing pre-and post-disturbance variables and compositions (Norden et al. 2009; O'Dea and Whittaker 2007; Moretti et al. 2006). Our project used similar methods to compare the characteristics of bird communities on peatland and non-peat lowland by focusing on the following: (1) species richness; (2) feeding guilds; and (3) response of avifauna to disturbance. We analyzed how bird populations responded to rapidly changing patterns of peatland land use. We compared data of bird populations inhabiting several different peatland landscapes in Riau, with those inhabiting non-peat lowland landscapes in Sumatra Island. We then discussed the vulnerability and potential resilience of the peat swamp ecosystem as indicated by avian popula-tion response to disturbance.

3.2 Methods

3.2.1 Datasets in Peatland

We used bird population datasets published by Fujita et al. (2016). The study site was located in the Bukit Batu area of the Giam Siak Kecil-Bukit Batu Biosphere Reserve, Riau Province, Indonesia (Fig. 3.1). Observations took place in the follow-ing sites: (1) natural peat swamp forest in the Bukit Batu Wildlife Reserve (hereafter referred to as BNF), which is managed by the Forestry Department of Riau; (2) plantation forest of *Acacia crassicarpa*, which is managed by the companies PT. Bukit Batu Hutani Alam and PT. Sakato Pratama Makmur (referred to as BAF);[1] (3) planted rubber (*Hevea brasiliensis*) forests, which are managed by local people (referred to as BRF); and (4) the residential village area in the Bukit Batu area (referred to as BRV; Fig. 3.2). BNF is dominated by *Palaquium sumatranum*, *Eugenia paludosa, Diospyros hermaphroditica, Calophyllum lowii*, and *Shorea teysmanniana*, with a peat depth of up to 6 m (Gunawan et al. 2012). According to Fujita et al. (2016), a bird survey was conducted in selectively logged forests which maintain a complex and multistory forest structure. Planted acacia forests are pure plantations of *Acacia crassicarpa* that are established on peatland where canals were dug in grids for drainage and transport purposes. Acacia trees are harvested within

---

[1] "PT." is the abbreviation for *Perseroan Terbatas*, and it means Limited Liability Company.

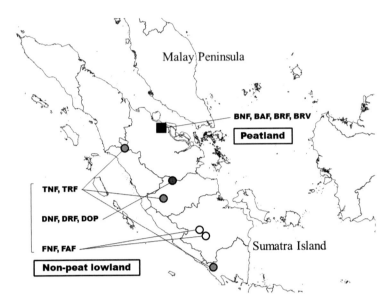

Fig. 3.1 Study sites. Abbreviations of study sites on the peatland of Fujita et al. (2016) are as follows: BNF natural peat swamp forest, BAF planted acacia forest, BRF planted rubber forest, BRV residential village. Study sites on non-peat lowland of Thiollay (1995) are TNF natural forest, TRF planted rubber forest, of Fujita et al. (2014) are FNF natural forest; FAF planted acacia forest, of Danielsen and Heegaard (1995) are DNF natural forest, DRF planted rubber forest, DOP oil palm plantation

5–6 years. The Acacia forest monocrop structure is simple—*Acacia crassicarpa* comprises the canopy layer, with only some ferns on the forest floor. Planted rubber forests, meanwhile, often include dense understories including shrubs, small trees, and saplings of native forest species. Although planted rubber forests range from highly-managed plantations with less understory to extensive forests with more understory trees, they are usually more diversified than industrial plantations. Land use in the residential village area consists of home gardens, rubber plantations, paddy fields, oil palm plantations, secondary forests (mainly *Macaranga* trees), coconut plantations, and burnt brush land.

Bird surveys were conducted using fixed-radius point count methods in twelve 1 km lines. Three surveys were conducted in each of the four land use categories. In each survey, four points were established at 250 m intervals, and every bird species that was seen or heard within a 25 m radius was recorded during the 20-min observation period. Observations were conducted in the morning and evening in March, May, and October 2011. In total, 34 censuses were made in the natural peat swamp forest (BNF), and 48 censuses were made in other sites (BAF, BRF, and BRV).

Fig. 3.2 Typical land use in peatland in Bukit Batu, Riau: (**a**) planted acacia forest (BAF); (**b**) residential village area (BRV); and (**c**) planted rubber forest (BRF) (Photos taken by Fujita)

3.2.2   Datasets in Non-peat Lowland

We selected three papers that surveyed the bird community in non-peat lowland in Sumatra (Table 3.1). The criteria for the paper selection were that the datasets must be from direct observations of birds using either the line-transect or point-count method, both of which are useful in analysis of bird community composition. Thiollay (1995) investigated bird fauna in lowland dipterocarp forests (hereafter referred to as TNF) and planted rubber forests (hereafter referred to as TRF) in the Jambi and West Sumatra provinces in May–June 1991 and July 1992, respectively (Fig. 3.1). The emergent trees (44–55 m) in TNF were mostly *Dipterocarpus* sp., *Shorea* sp., and *Cesalpiniaceae*, with canopy trees (35–45 m), including numerous *Dipterocarpaceae*, *Myristicaceae*, *Sapotaceae,* and fig trees (*Ficus* sp.). TRF consists mainly of para-rubber trees, *Hevea brasiliensis*, which are lower (closed canopy at 20–30 m) and have dense undergrowth. An average stand includes 750 trees per hectare, 65% of which are rubber trees mixed with 12–20 other species of forest and fruit trees. A line-transect survey was performed, where a sample was defined as the first 50 individual birds that were recorded within 20–30 m on either side of the transect. Twenty eight-samples were collected in both TNF and TRF. Since the

Table 3.1 Site description

Source	Fujita et al. (2016)	Thiollay (1995)	Danielsen and Heegaard (1995)	Fujita et al. (2014)
Soil	Peat land	Non-peat lowland	Non-peat lowland	Non-peat lowland
Location (province)	Riau	Jambi/West Sumatra	Riau/Jambi	South Sumatra
Surveyed year	2011	1991–1992	1991	2007–2008
Surveyed habitat type	Natural peat swamp forest, planted acacia forest, planted rubber forest, residential village area	Lowland dipterocarp forest, planted rubber forest	Lowland dipterocarp forest, planted rubber forest, oil palm plantation	Logged forest, planted acacia forest
Bird survey method	Fixed-radius point count	Line transect	Variable-distance line transect, mist netting	Fixed-radius point count

accessible tracts of the primary forest were limited, all the data from the three different areas were pooled into a single 28-sample set that was assumed to be representative of the western Sumatra natural forest. Counts were made from dawn to 11:00 h and from 15:00 h to sunset outside rainy periods.

Danielsen and Heegaard (1995) investigated bird fauna in both primary and logged lowland dipterocarp forests (hereafter referred to as DNF), planted rubber forests (hereafter referred to as DRF), and oil palm plantations (hereafter referred to as DOP) in Bukit Tigapuluh, Riau, and Jambi province in July–September 1991 (Fig. 3.1). The most prevalent tree families in DNF were *Euphorbiaceae*, *Lauraceae*, *Burseraceae*, and *Dipteroearpaceae*, which comprise a multistory forest structure, with a 26–38 m canopy layer, 18–26 m middle layer, and 10–21 m lower layer. The height of the canopy in DRF was 15–20 m, with emergents reaching 25 m, and 36% of the trees over 10 cm diameter reaching breast height (DBH) were estimated to be para-rubber tree (*Hevea brasiliensis*). Undergrowth consisted of 1–2 m tall perennials and thickets of bamboo *Dendrocalamus* sp.

The DOP plantation was 10,000 ha and consisted of mature oil palms, *Elaeis guineensis*. The average height of the palms was 11 m, and the density was 162 trees per ha. Within 1.5 m of the palms, an area was covered by dead palm leaves and ferns or no vegetation. The authors used a variable-distance line-transect method supplemented by systematic mist-netting. A line-transect survey route of 2 km was cut in a straight line in each habitat every 250 m. The route was walked by one person with a speed of 250 m per 15 min, including brief stopovers. The route was surveyed 4 times per day. In DNF and DRF, the route was surveyed for 40 h, while in DOP, the route was only surveyed for 20 h. Mist netting was carried out with 15–20 nets, which were mainly set at a height of 0.5–3.0 m for three to four consecutive days. The effort was 10,200–15,500 net-meter hours in DNF and DRF and only 3300 net-meter hours in DOP.

The third paper used to establish the present dataset, by Fujita et al. (2014), examined bird fauna in a lowland *Acacia mangium* plantation (FAF) and remnant logged natural forest (FNF). The height of the acacia canopy in FAF for the 4-year-old planted acacia forest was 10–20 m with less understory (1–10 m) vegetation, and the forest floor was occasionally covered by ferns or sparse scrub. The forest floor of the 1-year-old planted acacia forest had no vegetation other than 3–5 m of planted *A. mangium*. Most of the points in FNF consisted of trees that were 20–30 m tall and scattered with multistory vegetation. The other two sites of conserved logged forests (CF-07, CF-08) were burnt during a forest fire, and no living tall trees could be seen; instead, all sites were covered by dense bush.

Fixed-radius point counts were conducted in October–December 2007 and July–August 2008. Twenty survey points for FAF in the 1- and 4-year-old planted acacia forests and 12 survey points for FNF were selected. At each point, the bird species that were observed within a 25 m radius were recorded in the 10-min census. The number of censuses totaled 303 in FAF and 197 in FNF.

3.2.3 Feeding Guild

To compare and evaluate bird communities and forest ecosystem in terms of ecological function, feeding guilds of birds were used in the following analysis. A feeding guild is defined as the groups of species that have similar feeding habits. The following 17 feeding guild categories were used for the analysis: A-F, arboreal frugivore; AFG-I, arboreal foliage gleaning insectivore; AFG-IF, arboreal foliage gleaning insectivore–frugivore; A-FP, arboreal frugivore–predator; A-I, aerial insectivore; BG-I, bark gleaning insectivore; M-IP, miscellaneous insectivore–piscivore; NF, nectarivore-frugivore; NI, nectarivore-insectivore; NIF, nectarivore-insectivore-frugivore; R, raptor; S-I, sallying insectivore; SSG-I, sallying substrate gleaning insectivore; T-F, terrestrial frugivore; T-G, terrestrial granivore; T-I, terrestrial insectivore; and T-IF, terrestrial insectivore–frugivore. Each species was categorized in one of the feeding guilds, with reference to Lambert (1992), Styring et al. (2011), and Smythies (1999). Species previously unreferred to a feeding guild were categorized according to the feeding guild of the nearest species in the same genus.

3.2.4 Statistical Analysis

To compare the species richness at each site, all datasets were merged, and abundance data was converted to presence-absence data. Water birds, night birds, and swifts (excluding tree swifts) were omitted from the list. Scientific names followed the nomenclature of Sibley and Monroe (1990). Since observation efforts differ among sites, a direct comparison of species richness or number must be treated with caution. Therefore, we used the nonparametric Jackknife estimator of species richness to compare bird diversity among the sites. Jackknife 1 was calculated for the data of Bukit Batu and Fujita et al. (2014) using the software EstimateS ver 9.1 (Colwell 2013). Jackknife 1 was used since it has been widely used for years, including by Thiollay (1995). In this manner it was possible to compare data from at least three sites, with the exception of the dataset of Danielsen and Heegaard (1995).

To compare the species composition among the study sites, the ordination of sites and habitat types was performed against the presence-absence data. This was achieved using nonmetric multidimensional scaling (NMDS) with the package "vegan" (Oksanen et al. 2015) in the statistical software R version 3.2.2 (R Core Team 2015). NMDS was performed with Bray–Curtis as a dissimilarity index, and a maximum 100 times of iteration was conducted to obtain better ordination scores. Additionally, the species score against each axis was calculated, although only abundant species appeared with a label when several species appeared in the small area.

3.3 Results and Discussion

3.3.1 Bird Species Diversity on Peatland and Non-peat Lowland

The overall species number in the forests on peatland was lower than that of forests on non-peat lowland. The total number of species in peatland in the Bukit Batu area was 44 in the natural peat swamp forest (BNF), 20 in the planted acacia forest (BAF), 38 in the planted rubber forest (BRF), and 51 in the residential village area (BRV). In the forests in the non-peat lowland of Thiollay (1995), the number of species was 180 in natural forest (TNF) and 104 in the planted rubber forest (TRF). In the study by Danielsen and Heegaard (1995), there were 147 in the natural forest (DNF), 90 in the planted rubber forest (DRF), and 24 in the oil palm plantation (DOP). In the study by Fujita et al. (2014), there were 82 in the natural forest (FNF) and 63 in the planted acacia forest (FAF).

Out of 283 bird species in total, 101 species were recorded in forests on peatland in Bukit Batu, whereas in forests on non-peat lowland, 267 species were reported, and 85 species were seen in both peatland and non-peat lowland. Out of the 267 species that were seen in the forests on non-peat lowland, 182 species were not recorded in the forests on peatland. However, out of the 101 species on peatland, 16 species were seen only on peatland, of which four species (*Prionochilus thoracicus, Ficedula zanthopygia, Malacocincla abbotti,* and *Nectarinia calcosthetha*) were specifically seen only in natural forests on peatland (BNF). Of these, *P. thoracicus* and *N. calcosthetha* favor swamps or coastal forests as their

Table 3.2 Bird species in common in natural forests in Bukit Batu and other studies. Numbers in *italics* indicate the number of species in common, whereas the numbers in parentheses indicate Jaccard's coefficient of community (CC)

Number of species in common	Peatland	Non-peat lowland		
	Fujita et al. (2016)	Thiollay (1995)	Danielsen and Heegaard (1995)	Fujita et al. (2014)
(Jaccard's CC[a])	BNF	TNF	DNF	FNF
BNF	–	*32*	*37*	*25*
TNF	(0.167)	–	*108*	*51*
DNF	(0.240)	(0.493)	–	*54*
FNF	(0.248)	(0.242)	(0.309)	–
Total species number	44	180	147	82

[a]Jaccard's CC was calculated using the following formula: $CC = c/(a + b - c)$, where a is the number of species in site A, b is the number of species in site B, and c is the number of common species between site A and B. A larger number indicates a higher degree of similarity between the two sites

habitat. Additionally, we observed several species that prefer swamps, including *Malacopteron albogulare* and *Setornis criniger* (Fujita et al. 2012), although they are not listed in the point-count survey by Fujita et al. (2016). Gaither Jr (1994) reported that several bird species, including *M. albogulare,* were found in higher density in peat swamp forests compared to non-peat lowland forests. Although the species richness is low compared to forests on non-peat lowland, forests on peatland could be home to several bird species that favor a swampy environment or species that avoid the highly competitive pressure in forests on non-peat lowland.

Jaccard's coefficient of community (CC) index, which represents the community similarity between two sites, indicates that the peatland bird community has a low similarity (0.167–0.248) with other non-peat lowland bird communities, whereas the non-peat lowland bird community has a higher similarity (0.242–0.493) among sites (Table 3.2).

Since the efforts and methods of observation differ among sites, it is not possible to compare species richness directly. The Jackknife estimator of species richness, the Jack 1 mean, indicates the true species richness of natural forest in Bukit Batu (BNF) as 77.2, which is 33 species greater than the observed species richness (Table 3.3). Compared to natural forests on non-peat lowland, of which the estimated species richness is 241 species in TNF and 114.8 in FNF, natural forests on peatland have a lower species richness.

These measurements indicate poor species diversity in peat swamp forests, which follows the hypothesis that peat swamp forests support depauperate animal communities due to the cascading influence of poor soils (Janzen 1974). Nevertheless, the peat swamp forest is a unique ecosystem supporting several bird species that are not present in other ecosystems, including forests on non-peat lowland.

These direct comparisons of landcover and bird populations must be considered with caution for several reasons. First, observation efforts differ significantly in the

Table 3.3 Comparison of species richness and estimators of natural forest among four sites. The proportion of observed richness to the computed estimates is shown in parentheses

	Peatland	Non-peat lowland		
	Fujita et al. (2016)	Thiollay (1995)	Danielsen and Heegaard (1995)	Fujita et al. (2014)
	BNF	TNF	DNF	FNF
Observed species richness	44	180	147	82
Observed individuals	155	1400	–	823
Jack 1 mean: Jackknife estimate of species richness	77.2 (60.0)	241 (74.7)	–	114.8 (71.4)
Efforts				
Line-transect (hours)	–	28–140[a]	40	–
Point-census (hours)	11	–	–	33
Mist-net (net-hour)	–	–	10,200–15,500	–

[a]A sample is defined as the first 50 individual birds that were recorded along a random footpath within 20–30 m on either side of the transect. Fifty individual samples were accumulated within 1–4 h along a track

different studies—up to 10 times in observation hours (Table 3.3). The lower observation time in the natural forest on peatland is due to limited accessibility, time, and labor power, all of which may be linked to the lower number of species observed. Comparisons of estimators of species richness, such as Jackknife, have solved this problem to some extent. Second, the observation methods differed among sites and included line transect, point count, and mist netting. Even the line-transect method differed slightly among sites, as Thiollay (1995) counted 50 individuals from the beginning, with no regard to the length of the line, whereas Danielsen and Heegaard (1995) walked the 250 m line and counted all birds observed. Third, habitat complexity differed among sites, as Thiollay (1995) merged all data from several patches from different regions and treated it as a single dataset for the primary forest. The regions include hill forests, which resulted in the presence of montane species and increased the overall species richness. Moreover, regarding Bukit Batu, all study points of natural peat swamp forest (BNF) were selectively logged a few years before the survey, perhaps affecting the study results. In short, such discrepancies in datasets indicates that the hypothesis that peatlands support poor bird species diversity still requires further intensive research.

3.3.2   Feeding Guild Change in Disturbed Sites on Peatland

In the disturbed sites of peatland in Bukit Batu, namely planted acacia forest (BAF), planted rubber forest (BRF), and residential village (BRV), a decline in the sallying insectivores (S-I and SSG-I) was observed (Table 3.4). This decline was also observed in non-peat lowland forest. BRV was characterized by more granivores (T-G) and terrestrial doves (T-IF), which reflect the abundance of open spaces and

Table 3.4 Number of species for each feeding guild

| Feeding guild | | | | | Peatland | | | | Non-peat lowland | | | | | | |
| | | | | | Fujita et al. (2016) | | | | Thiollay (1995) | | Danielsen and Heegaard (1995) | | | Fujita et al. (2014) | |
Category	Fruit/grain	Nectar	Insect	Fish/animal	Natural forest BNF	Planted acacia forest BAF	Rubber jungle BRF	Residential village BRV	Natural forest TNF	Rubber jungle TRF	Natural forest DNF	Rubber jungle DRF	Oil palm plantation DOP	Natural forest FNF	Planted acacia forest FAF
A-F	+				1	0	4	3	14	6	11	6	2	3	4
AFG-I			+		16	5	13	11	40	32	34	25	7	21	15
AFG-IF	+		+		7	6	5	6	26	20	21	16	2	14	10
A-FP	+			+	0	1	1	3	9	3	8	5	0	1	1
A-I			+		0	0	0	0	0	0	1	1	1	0	2
BG-I			+		1	0	1	3	10	3	10	2	0	2	3
M-IP			+	+	0	3	1	3	4	1	5	5	2	1	1
M-P				+	0	1	0	0	0	0	0	0	0	0	0
NF	+	+			1	0	0	1	1	1	1	1	0	1	1
NI		+	+		2	0	1	1	9	7	4	4	0	8	4
NIF	+	+	+		5	1	5	4	12	9	7	5	0	6	3
R				+	1	1	1	2	6	3	6	0	0	3	2
S-I			+		5	1	3	3	15	8	11	5	2	8	9
SSG-I			+		5	0	1	1	16	6	11	5	0	7	4
T-F	+				0	0	0	0	1	0	2	1	0	1	1
T-G	+				0	0	0	6	0	1	1	2	6	1	1
T-I			+		0	1	2	2	13	4	10	6	1	4	1
T-IF	+		+		0	0	0	2	4	0	4	1	1	1	1
Total number of species					44	20	38	51	180	104	147	90	24	82	63

grass seeds as their food (Fig. 3.2b). Hornbills (A-FP) appeared more frequently in the residential village than expected. This could be because they were more easily detectable in open spaces than in the forest, since their sound was heard more often in the forested sites than BNF (although not every species was recorded when their call was beyond our observation range of 25 m).

Compared to natural forests on non-peat lowland (TNF, DNF, and FNF), natural forest in Bukit Batu (BNF) was characterized by fewer terrestrial insectivores (T-I), such as *Trichastoma* sp., *Enicurus* sp., *Pitta* sp., and *Napothera* sp., and fewer woodpeckers (BG-I) (Table 3.4). Gaither Jr (1994) reported similar results in a mist-net survey of understory birds in peat swamp forests in West Kalimantan. He found litter-gleaning insectivores (T-I in our category), tree foliage-gleaning insectivores (AFG-I in our category), and bark-gleaning insectivores in significantly lower numbers in the peat swamp forest than in the dipterocarp forest. Posa (2011) had partly contradicting results in the peat swamp forest in Central Kalimantan, as six species (8%) of woodpeckers (BG-I) were recorded in peat swamp forests, although litter-gleaning insectivores (T-I in our category) only comprised three species (4%). The abundance of woodpeckers was associated with the number of dead trees in the degraded peat swamp forest due to recent drought stress (Posa 2011). The decline in TI and BGI in Bukit Batu might be the result of logging activity in the natural forest. As recently as 2009, illegal logging has occurred in most of the accessible areas, including our study plots.

The low number of arboreal frugivores (A-F) could be explained by the seasonal change. Gaither Jr (1994) reported that many frugivore species were observed in June in the peat swamp forest, but not in other months.

3.3.3 Shift in Bird Community Composition along the Disturbance

Many fewer species were found in disturbed sites than in the natural peatland forests in Bukit Batu. Out of the 44 species found in natural forests in Bukit Batu (BNF), only eight species in planted acacia forest (BAF), 12 species in the planted rubber forest (BRF), and nine species in residential village (BRV) were shared, which indicated that more than 70% of species were lost in the disturbed sites. However, in non-peat lowland, 88 species in the planted rubber forest (TRF) out of 180 species in the natural forest (TNF), 66 species in the planted rubber forest (DRF) out of 147 species in natural forest (DNF), and 42 species in the planted acacia forest (FAF) out of 82 species in natural forest (FNF) were found, which account for approximately 50% of the species loss in the disturbed sites. The oil palm plantation showed a drastic species loss of up to 94%, as only eight species were found in DOP, in comparison to 147 species in the natural forest (DNF).

The NMDS analysis showed that the bird community composition in forests on peatland is unique compared to forests on non-peat lowland (Fig. 3.3a). Points of

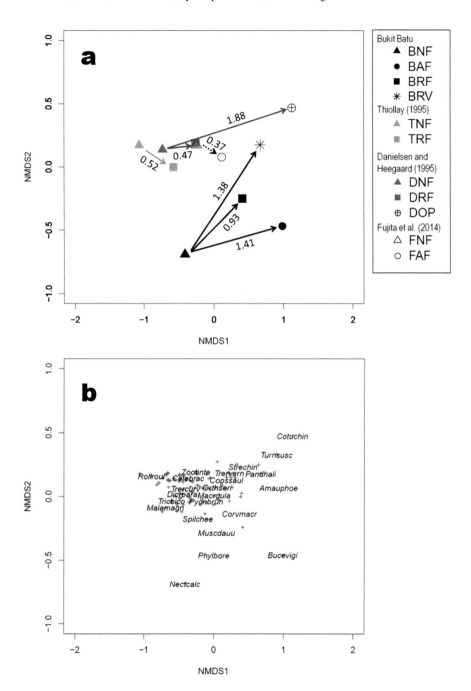

Fig. 3.3 Bird community similarities using an NMDS ordination of study points (**a**) and species (**b**). (**a**) Each point corresponds to a study point. Abbreviations in the legends are as follows: BNF natural peat swamp forest in Bukit Batu (closed triangle), BAF planted acacia forest in Bukit Batu (closed circle), BRF planted rubber forest in Bukit Batu (closed square), BRV residential village in Bukit Batu (star), TNF natural forest of Thiollay 1995 (pale gray triangle), TRF planted rubber of

non-peat lowland (TNF, TRF, DNF, DRF, DOP, FNF, and FAF) showed a higher NMDS2 value than points of peatland, except for BRV, which is BNF, BAF, and BRF. This indicates that peatland has different avifauna compared to the forests on non-peat lowland. BRV showed a similar NMDS2 value as that of non-peat lowland; this might have resulted from either the high modification to the village or the presence of non-peat soil in the villages, which are located on the coastline. The NMDS2 axis was negatively correlated with the presence of *Nectarinia calcosthetha*, which is restricted to peat swamp forests (Fig. 3.3b).

While the NMDS2 axis divided forests on non-peat lowland and peatland, NMDS1 axes showed the degree of disturbance. On both peatland and non-peat lowland, the natural forest points (BNF, TNF, DNF, and FNF) showed lower NMDS1 values than disturbed points, such as planted acacia forest, planted rubber forest, oil palm plantation, and residential village (Fig. 3.3a, direction of disturbance shown in the arrows for each site). Species that had a positive correlation with the NMDS1 axis include *Cotumix chinensis* and *Turnix suscitator*, both of which are open space species, and species that had a negative correlation include *Calyptomena viridis* and *Trichastoma bicolor*, both of which are species inhabiting less disturbed natural forests (Fig. 3.3b).

Disturbed points on peatland in Bukit Batu (BAF, BRF) showed a higher NMDS1 value and NMDS2 value compared to natural forests (BNF). This indicates that when natural forests are converted to planted acacia and rubber forests bird communities change accordingly, shifting toward the avifauna of non-peat lowlands. For example, *Nectarinia calcostetha*, which is found in natural peatland forests, has not been recorded in other habitat types. Disturbance of peatland therefore will likely lead to the homogenization of avifauna and loss of uniqueness, which would then lead to a loss in overall biodiversity.

Peatland-based bird community composition responded sensitively to land use change. Euclidean distances of the NMDS values of community composition between natural forest and planted rubber forest in non-peat lowland were 0.52 in TNF-TRF, 0.47 in DNF-DRF, and 0.93 in BNF-BRF in peatland, indicating a

---

Fig. 3.3 (continued) Thiollay 1995 (pale gray square), FNF natural forest of Fujita et al. 2014 (open triangle), FAF planted acacia forest of Fujita et al. 2014 (open circle), DNF natural forest of Danielsen and Heegaard 1995 (dark gray triangle), DRF planted rubber forest of Danielsen and Heegaard 1995 (dark gray square), DOP oil palm plantation of Danielsen and Heegaard 1995 (cross in open circle). The arrows indicate the directions from the natural forest to the disturbed habitat at each site. The values near the arrows indicate the Euclidean distances calculated from the NMDS 1 and 2 values. (**b**) In the species plot, only abundant species are labeled where several species are clustered. Abbreviations for the species names are as follows: *Amauphoe Amaurornis phoenicurus, Bucevigi Buceros vigil, Celebrac Celeus brachyurus, Copssaul Copsychus saularis, Corvmacr Corvus macrorhynchos, Cotuchin Cotumix chinensis, Dicrpara Dicrurus paradiseus, Macrgula Macronous gularis, Malamagn Malacopteron magnum, Muscdauu Muscicapa dauurica, Nectcalc Nectarinia calcostetha, Orthseri Orthotomus sericeus, Pandhali Pandion haliaetus, Phylbore Phylloscopus borealis, Pycnbrun Pycnonotus brunneus, Rollroul Rollulus rouloul, Spilchee Spilornis cheela, Strechin Streptopelia chinensis, Trercurv Treron curvirostra, Trervern Treron vernans, Trichico Trichastoma bicolor, Turnsusc Turnix suscitator, Zootinte Zoothera interpres*

1.7–1.9-times longer distance in peatland (Fig. 3.3a). As distances on NMDS values indicate differences in community composition, a longer Euclidean distance indicates a larger difference in community composition. Euclidean distances between natural forest and planted acacia forest in non-peat lowland were 0.37 in FNF-FAF and 1.41 in BNF-BAF in peatland, which shows a 3.8-times longer distance in peatland (Fig. 3.3a). Although avifauna clearly responded more strongly to disturbance in peatland, the reason for this vulnerability is unclear. It may be attributed low biodiversity and the idiosyncratic characteristics of the peat swamp ecosystem. To verify the resilience in the peat swamp forest, repetitive data from long-term studies on community composition and analyses from the perspective of landscape ecology are necessary. For example, differences in landscape structure and scale in peatlands might have significant impacts on bird communities, since plantations are often developed on a large scale due to the flat topography and lower utilization by local people.

3.3.4 Conclusion: Peatland Ecosystem for Birds

Our survey and analysis lead us to conclude the following: (1) although the peatland ecosystem has a comparably poorer species diversity than non-peat lowland forest, peatlands contains unique and idiosyncratic avifauna; (2) peat swamp forest avifauna lack terrestrial insectivore and woodpeckers; and (3) avifauna in peatland are sensitive to forest conversion to planted tree forest, and can experience a loss of more than 70% of species. In other words, unique and idiosyncratic avifauna on peatland are only maintained by natural peat swamp forest. Our first two findings support those of Janzen (1974), Gaither Jr (1994), and Posa (2011). The third finding, which has the most stimulating implication, suggests that the peat swamp ecosystem has a higher vulnerability to disturbance compared to non-peat lowland. It is still in the stage of hypothesis. Peat swamp vulnerability may be due to poor biodiversity and the idiosyncratic characteristic of the peat swamp ecosystem. Long-term studies on bird community and landscape ecology are required to evaluate this hypothesis. Nevertheless, our study found several important implications for peatland management, considering that there are limited studies on the effect of peatland disturbance on biodiversity.

Samejima et al. (2016) listed several other bird species detected by camera trap. *Lophura erythrophthalma* and *Melanoperdix niger* were both vulnerable species, and they were both restricted to natural peat swamp forest and not observed in the planted acacia forest. This indicates that disturbance has negative effects on these forest-dependent species, and the implication is consistent with our results. To conserve the unique peat swamp avifauna and maintain the ecosystem services that are provided by biodiversity, peat swamp forest should not be converted to plantations and agricultural lands, as peatland ecosystems have low resilience to disturbance.

Acknowledgments   This study was partly supported by the JSPS Global COE Program "In Search of Sustainable Humanosphere in Asia."

References

Anderson J (1961) The ecology and forest type of the peat swamp forests of Sarawak and Brunei in relation to their silviculture. Forest Department Sarawak, Kuching

Barlow J, Gardner TA, Araujo IS et al (2007) Quantifying the biodiversity value of tropical primary, secondary, and plantation forests. PNAS 104:18555–18560. https://doi.org/10.1073/pnas. 0703333104

Bruenig E (1990) Oligotrophic forested wetlands in Borneo. In: Lugo AE, Brinson M, Brown S et al (eds) Ecosystems of the world, Forested wetlands, vol 15. Elsevier, Amsterdam, pp 299–334

Bruenig E, Droste H (1995) Structure, dynamics and management of rainforests on nutrient-deficient soils in Sarawak. In: Primack RB, Lovejoy TE (eds) Ecology, conservation and management of Southeast Asian rainforests. Yale University Press, New Haven, pp 41–53

Cheyne SM, Macdonald DW (2011) Wild felid diversity and activity patterns in Sabangau peat-swamp forest, Indonesian Borneo. Oryx 45:119–124

Cheyne SM, Husson SJ, Chadwick RJ et al (2010) Diversity and activity of small carnivores of the Sabangau Peat-swamp Forest, Indonesian Borneo. Small Carniv Conser 43:1–7

Colwell RK (2013) EstimateS: statistical estimation of species richness and shared species from samples. Version 9. User's guide and application. http://purl.oclc.org/estimates. Accessed 8 Aug 2016

Corlett RT (2009) The ecology of tropical East Asia. Oxford University Press, Oxford

Danielsen F, Heegaard M (1995) The birds of Bukit Tigapuluh, Southern Riau, Sumatra. KUKILA 7:99–120

Elmqvist T, Folke C, Nystrom M et al (2003) Response diversity, ecosystem change, and resilience. Front Ecol Environ 1:488–494. https://doi.org/10.1890/1540-9295(2003)001[0488:RDECAR] 2.0.CO;2

Fujita MS, Irham M, Fitriana YS et al (2012) Mammals and birds in Bukit Batu area of Giam Siak Kecil–Bukit Batu Biosphere Reserve, Riau, Indonesia. Kyoto Working Papers on Area Studies No. 128 (G-COE Series 126), pp 1–70. Center for Southeast Asian Studies, Kyoto

Fujita MS, Prawiradilaga D, Yoshimura T (2014) Roles of fragmented and logged forests for bird communities in industrial *Acacia mangium* plantations in Indonesia. Ecol Res 29:741–755. https://doi.org/10.1007/s11284-014-1166-x

Fujita MS, Samejima H, Haryadi DS et al (2016) Low conservation value of converted habitat for avifauna in tropical peatland on Sumatra, Indonesia. Ecol Res 31:275–285. https://doi.org/10. 1007/s11284-016-1334-2

Gaither JC Jr (1994) Understory avifauna of a Bornean peat swamp forest: is it depauperate? Wilson Bull 106:381–390

Gunawan H, Kobayashi S, Mizuno K et al (2012) Peat swamp forest types and their regeneration in Giam Siak Kecil-Bukit Batu Biosphere Reserve, Riau, East Sumatra, Indonesia. Mires Peat 10: 1–17

Janzen DH (1974) Tropical blackwater rivers, animals, and mast fruiting by Dipterocarpaceae. Biotropica 6:69–103. https://doi.org/10.2307/2989823

Koh LP, Miettinen J, Liew SC et al (2011) Remotely sensed evidence of tropical peatland conversion to oil palm. PNAS 108:5127–5132. https://doi.org/10.1073/pnas.1018776108

Lambert FR (1992) The consequences of selective logging for Bornean lowland forest birds. Philos Trans R Soc B 335:443–457. https://doi.org/10.1098/rstb.1992.0036

Laumonier Y, Uryu Y, Stüwe M et al (2010) Eco-floristic sectors and deforestation threats in Sumatra: identifying new conservation area network priorities for ecosystem-based land use planning. Biodivers Conserv 19:1153–1174. https://doi.org/10.1007/s10531-010-9784-2

Millennium Ecosystem Assessment (2005) Ecosystems and human well-being: synthesis. Island Press, Washington, DC

Moretti M, Duelli P, Obrist MK (2006) Biodiversity and resilience of arthropod communities after fire disturbance in temperate forests. Oecologia 149:312–327. https://doi.org/10.1007/s00442-006-0450-z

Norden N, Chazdon RL, Chao A et al (2009) Resilience of tropical rain forests: tree community reassembly in secondary forests. Ecol Lett 12:385–394. https://doi.org/10.1111/j.1461-0248.2009.01292.x

O'Connell TJ, Jackson LE, Brooks RP (2000) Bird guilds as indicators of ecological condition in the Central Appalachians. Ecol Appl 10:1706–1721. https://doi.org/10.1890/1051-0761(2000)010[1706:BGAIOE]2.0.CO;2

O'Dea N, Whittaker RJ (2007) How resilient are Andean montane forest bird communities to habitat degradation? Biodivers Conserv 16:1131–1159. https://doi.org/10.1007/s10531-006-9095-9

Oksanen J, Blanchet FG, Kindt R et al (2015) Vegan: community ecology package. R package version 23-0 http://CRANR-projectorg/package=vegan. Accessed 24 Jun 2015

Philips VD (1998) Peatswamp ecology and sustainable development in Borneo. Biodivers Conserv 7:651–671. https://doi.org/10.1023/A:1008808519096

Posa MRC (2011) Peat swamp forest avifauna of Central Kalimantan, Indonesia: effects of habitat loss and degradation. Biol Conserv 144:2548–2556. https://doi.org/10.1016/j.biocon.2011.07.015

Posa MRC, Wijedasa LS, Corlett RT (2011) Biodiversity and conservation of tropical peat swamp forests. Bioscience 61:49–57. https://doi.org/10.1525/bio.2011.61.1.10

R Core Team (2015) R: a language and environment for statistical computing. R foundation for statistical computing, Vienna. https://www.R-project.org/. Accessed 10 July 2015

Samejima H, Fujita MS, Muhammad A (2016) Biodiversity in peat swamp forest and plantations. In: Mizuno K, Fujita MS, Kawai S (eds) Catastrophe and regeneration in Indonesia's peatlands: ecology, economy and society, Kyoto CSEAS series on Asian studies, vol 15. NUS Press/Kyoto University Press, Singapore/Kyoto, pp 353–379

Şekercioğlu CH, Wenny DG, Whelan CJ (2016) Why birds matter: avian ecological function and ecosystem services. University of Chicago Press, Chicago and London

Sibley CG, Monroe BL (1990) Distribution and taxonomy of birds of the world. Yale University Press, New Haven

Smythies BE (1999) The birds of Borneo, 4th edn. Natural History Publications (Borneo), Kota Kinabalu

Styring AR, Ragai R, Unggang J et al (2011) Bird community assembly in Bornean industrial tree plantations: effects of forest age and structure. For Ecol Manag 261:531–544. https://doi.org/10.1016/j.foreco.2010.11.003

Thiollay JM (1995) The role of traditional agroforests in the conservation of rain forest bird diversity in Sumatra. Conserv Biol 9:335–353

Uryu Y, Mott C, Foead N et al (2008) Deforestation, forest degradation, biodiversity loss and CO_2 emissions in Riau, Sumatra, Indonesia: one Indonesian province's forest and peat soil carbon loss over a quarter century and its plans for the future. WWF Indonesia technical report, WWF Indonesia, Jakarta

Whitemore TC (1984) Tropical rain forest of the Far East. Clarendon Press, Oxford

Whitten T, Damanik SJ, Anwar J et al (2000) The ecology of Sumatra. Periplus, Hong Kong

Yule CM (2010) Loss of biodiversity and ecosystem functioning in Indo-Malayan peat swamp forests. Biodivers Conserv 19:393–409

Chapter 4
Impact of Industrial Tree Plantation on Ground-Dwelling Mammals and Birds in a Peat Swamp Forest in Sumatra

Hiromitsu Samejima, Motoko S. Fujita, and Ahmad Muhammad

Abstract Peat swamp forests are one of the unique ecosystems of Southeast Asia. These forests are not only a large carbon stock, but also a refuge for rich biodiversity. To understand the faunal composition and the effect of land-use changes in peat swamp forests, we investigated ground-dwelling mammals and birds using camera traps in a natural peat swamp forest and acacia forests planted in two industrial tree plantations in the Giam Siak Kecil-Bukit Batu Biosphere Reserve, Riau, Indonesia, in the Island of Sumatra.

We obtained a total of 1856 records, comprising 23 species and including 11 vulnerable or endangered species, in ten plots. The range of mean trapping rates (number of records per 100 camera working days) of all animals in each plot in natural peat swamp forests were 9.22–51.85 (mean: 29.16) and 8.75–31.76 (16.42) in the wildlife reserve and protected area of the plantations respectively. The range in planted acacia forest was 2.29–6.38 (4.02). Few species were recorded in the planted acacia forests, and the species composition was different from that in the natural peat swamp forests. These differences indicate that conversion from natural peat swamp forests to planted acacia forest through development of industrial tree plantations resulting in decreased density and species richness of ground-dwelling mammals and birds. Because the ground-dwelling mammal and bird community in natural peat swamp forest is vulnerable to land use change, conservation of the remnant natural peat swamp forests and appropriate landscape design of industrial tree plantations are considered important to maintain the ecosystem.

Keywords Peatland · Sumatra · Industrial tree plantation · Camera trap · Mammal · Bird

H. Samejima (✉)
Institute for Global Environmental Strategies, Hayama, Kanagawa, Japan
e-mail: samejima@iges.or.jp

M. S. Fujita
Center for Southeast Asian Studies, Kyoto University, Kyoto, Japan

A. Muhammad
University of Riau, Pekanbaru, Riau, Indonesia

© The Author(s) 2023
K. Mizuno et al. (eds.), *Vulnerability and Transformation of Indonesian Peatlands*, Global Environmental Studies, https://doi.org/10.1007/978-981-99-0906-3_4

4.1 Introduction

Lowland peatlands are a unique ecosystem in Southeast Asia, distributed mostly in the east coast of Sumatra, Central and West Kalimantan, Sarawak, and southern Papua (Hooijer et al. 2010). Peat soil consists mainly of decomposed trunks, branches and roots of trees which have been sedimented over a period of 4000–26,000 years (Page et al. 1999, 2004). Peatlands are nutrient poor and have extremely acidic soil (pH 3.0–4.5) because they are situated away from large rivers that bring mineral nutrients downstream from mountainous areas. The total area of peatlands in Southeast Asia is estimated to be 24.8 million ha, comprising 56% of the global total. Southeast Asian peatlands are estimated to store 68.5 Gt of carbon, comprising 11–14% of the carbon stock in the global peatland (Page et al. 2011).

The natural vegetation of Southeast Asian peatlands is peat swamp forest (Anderson 1961; Bruenig 1990; Posa et al. 2011). Peat swamp forests in Southeast Asia function not only as a significant carbon sink, but also as a refuge for many endangered species in the region. The faunal composition of this vast ecosystem has not been as well studied as the floral composition, however (Gaither Jr 1994; Page et al. 1997; Whitemore 1984). Because of the poor nutrient content of the peat soil and the low primary productivity of the forests, they have been presumed to have lower diversity and abundance of animals than surrounding lowland forest (Janzen 1974; Posa et al. 2011; Whitten et al. 2000). Many endangered species, however, are known to inhabit peat swamp forests (Buckley et al. 2006; Cheyne et al. 2008; Felton et al. 2003; Morrogh-Bernard et al. 2003; Yule 2010). Furthermore, Ehlers Smith and Ehlers Smith (2013), Johnson et al. (2005), and Quinten et al. (2010), indicated that densities of several primate species, including orangutan, in peat swamp forests were higher than those in adjacent lowland forests. Gaither Jr (1994) also detected that some bird species were more abundant in peat swamp forest than in lowland forest.

Recently, ground-dwelling mammals and birds in natural peat swamp forest have been investigated with camera traps. Camera traps are automatic digital cameras equipped with infrared motion sensors that present new methods for investigation of faunal diversity (Cannon et al. 2007; Cheyne ct al. 2010; Cheyne and Macdonald 2011; Posa 2011; Kays and Slauson 2008; Rowcliffe and Carbone 2008; McCallum 2012; Burton et al. 2015; Nakashima et al. 2018). Mohd-Azlan (2004) conducted camera-trapping studies in a natural peat swamp forest at Maludam National Park in Sarawak; Cheyne et al. (2010) and Cheyne and Macdonald (2011) conducted the same in the Sebangau National Park in Central Kalimantan; and Posa (2011) along the Kapuas River, Central Kalimantan. However, some of those studies reported only felid and bird species, and the others ran their camera traps for only a limited number of days.

In recent decades, large-scale plantations of fast-growing trees and oil palm have been developed in the peatlands in Southeast Asia (Uryu et al. 2008; Corlett 2009; Yule 2010). Posa et al. (2011) estimated that a minimum of 64% of the historical peat swamp forest in Southeast Asia has been lost so far. The deforestation and drainage of the peatlands results in huge emission of carbon dioxide, which accelerates global warming (Couwenberg et al. 2009). Carbon dioxide emissions from peat swamp forest degradation in Southeast Asia during 1997–2006 was estimated to be as much as 0.30 Pg C year^{-1}, or 3% of total global anthropogenic carbon emissions (van der Werf et al. 2009).

Conversion of natural peat swamp forests to industrial tree plantations and oil palm plantations can also cause significant loss of biodiversity, but this phenomenon is not yet well studied (Posa 2011; Posa et al. 2011). Some animal species inhabiting peat swamp forests may be able to adapt to the new vegetation (Meijaard et al. 2010), but plantantions may also lead to serious loss of biodiversity, especially among species vulnerable to development in surrounding lowland forests in the region (Fitzherbert et al. 2008; McShea et al. 2009). Understanding the effects of plantation development on the animal community in peat swamp forests is necessary for proper management of the peat swamp landscape of the region.

To understand the faunal diversity of natural peat swamp forests and the impact of plantation development on these, we investigated species composition and abundance of ground-dwelling, medium to large mammals and birds using camera traps in natural and planted acacia forests in Giam Siak Kecil-Bukit Batu Biosphere Reserve (GSK-BB), Riau Province, Indonesia, in the Island of Sumatra. Some of the results were already published as Samejima et al. (2016).

4.2 Materials and Methods

4.2.1 Study Area

The Giam Siak Kecil-Bukit Batu Biosphere Reserve (GSK-BB) is located along the Strait of Malacca, covering an area of 7053 km^2. This biosphere reserve was established by a private-sector initiative and was registered by UNESCO's Man and Biosphere (MAB) program in 2009. GSK-BB consists of three zones: a core area (1787 km^2), buffer zone (2224 km^2), and transition area (3041 km^2). The core area consists natural peat swamp forest mostly in the Giam Siak Kecil Wildlife Reserve and the Bukit Batu Wildlife Reserve. The buffer zone comprises planted forests of fast-growing tree species (*Acacia crassicarpa* and *A. mangium*) inside four industrial tree plantations, called Hutan Tanaman Industri (HTI) in Indonesia. The plantations also have natural forests in areas adjacent to the Wildlife Reserves, which constitute a part of the core area. The natural forest was preserved in accordance of the Ministry of Forestry Decree No.70/Kpts-II/1995 and No.246/Kpts-II/1996, requiring that 10% of a concession area should be protected as "Kawasan Lindung." The outermost transition area of GSK-BB contains agricultural fields of smallholders and some oil palm plantations.

This study was conducted in the Bukit Batu area (BB) in the northeastern part of GSK-BB covering an area of approximately 800 km^2 (Fig. 4.1, Fujita et al. 2016). BB is located at latitude 1° 17–28′ north and longitude 101° 42′–102° 00′ east,

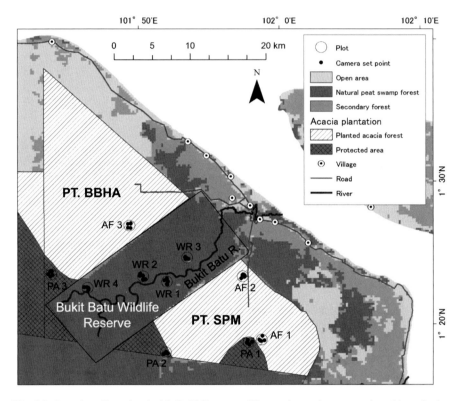

Fig. 4.1 Location of ten plots inside Bukit Batu area. Five random points were selected in each plot to set camera traps

including Bukit Batu Wildlife Reserve and industrial tree plantations of Bukit Batu Hutan Alam (PT. BBHA) and Sekato Pratama Makmur (PT. SPM). All of the study area comprises 4–8 meter-thick peat (Simbolon et al. 2011), with a water level of about 40–100 cm below ground (Kozan 2016).

The annual amount of rainfall in the study area is highly variable and its seasonality was unclear. Annual rainfall at Pekanbaru, located 100 km southwest of BB, ranged from 102 mm to 284 mm from 1999 to 2001, but was 1074–2596 mm from 2003 to 2011 (National Climatic Data Center 2012). Clear dry periods existed in some years, but the timing of such dry periods varied every year. Temperature ranged from 23 °C to 32 °C throughout the year.

The core zone in our study area was mostly covered by natural peat swamp forest. However, large trees within a few kilometers of both banks along the lower part of the Bukit Batu River in the Wildlife Reserve were logged by a logging company and smallholders during 1990s and 2000s (Watanabe et al. 2016). In the Wildlife Reserve, Gunawan et al. (2012) established two 0.5-hectare plots and measured all trees with diameter at breast height (DBH) > 3 cm. In the protected area, Partomihadjo et al. (2011) established two 1-ha plots and measured all trees with

DBH > 10 cm. Tree species composition in these four plots were similar, the common major species being *Palaquim sumatrana/dasyphyllum*, *Syszigium paludosa*, *Diospyros hermaphroditica*, *Mezzetia parvifolia*, and *Madhuca motleyana*. The forest was categorized as mixed swamp forest following the classification by Anderson (1976).

The buffer zone comprises planted forest of *A. crassicarpa* with networks of canals for the water management and transportation constructed at intervals of 500–3000 m. *A. crassicarpa* is a fast-growing species well adapted to peat soil. It grows in dense stands within 3 years of planting. The two plantations in our study area were established during 1999 (Watanabe et al. 2016) and developed until 2006. Acacia trees are harvested and replanted on a 5–6-year rotation, and these planted acacia forests consist of patches of even-aged trees. The forest floor of the planted acacia forest is dominated by *Dicranopteris* sp. and *Blechnum* sp.

Hunting pressure on medium to large mammals in this area is considered to be low because there are no residents inside the study area. Also, most of the nearby inhabitants are Muslim (Suzuki et al. 2016) who rarely hunt wild mammals. Nevertheless, we found snares to catch wild boar set by outsiders and the skull of a sambar deer (*Rusa unicolor*) at a camp used by poachers in the Wildlife Reserve. Cause of death of the sambar deer—either at the hands of poachers or natural causes—was unclear.

4.2.2 Camera Trapping

To evaluate the species composition and abundance of ground-dwelling mammals and birds, we deployed camera traps—automatic digital cameras with infrared motion sensors (Bushnell Trophy Cam, Model 119435, Bushnell, Olathe, KS), in BB, from December 2010 to October 2011.

We established four plots in the natural peat swamp forest inside the Bukit Batu Wildlife Reserve (WR1, WR2, WR3, WR4) and three plots in the natural peat swamp forest inside the protected areas of the two industrial tree plantations (PA1, PA2, PA3) (Fig. 4.1 and Table 4.1). While WR4 and PA1–PA3 were in primary forests, WR1–WR3 were in forests degraded by previous logging activities. WR1–WR4 were located about 500–1000 m from the banks of Batu River. The altitudes of WR1–WR3 were lower than those of all the other plots. PA1–PA3 were located away from the river, but located close to canals (approximately 500–1000 m) in the plantations. We also established three plots in the planted forests of *A. crassicarpa* inside the two plantations (AF1, AF2, AF3). They are all close to canals (approximately 500 m). Each of the ten plots covered a circular area with a radius of 500 m. The altitudes of these plots ranged from 8 m to 40 m above sea level. The forest floor was generally flat and covered in wooden debris. Ground surface of the plots were not flooded during the study period.

We set camera traps at five random points in each plot and compared the species compositions captured at each plot and a mean trapping rate (MTR) for each species.

Table 4.1 Longitude and latitude of centers of the ten plots

Plot ID	North	East	Altitude (m)	Status
WR 1	1.38292	101.8662	8–25	Wildlife Reserve (Natural forest)
WR2	1.38901	101.838	10–25	
WR3	1.41084	101.8904	15–25	
WR4	1.37579	101.7724	30–35	
PAl	1.31237	101.9635	15–25	Protected area of HTI (Natural forest)
PA2	1.30054	101.8649	30–40	
PA3	1.39133	101.7287	27–34	
AFl	1.31523	101.979	25–35	Planted acacia forest of HTI
AF2	1.38917	101.9562	20–30	
AF3	1.44831	101.8222	25–35	

The camera was mounted onto a tree at a height of 50–100 cm above ground. We directed the camera to focus downward to capture a field-of-view of approximately 2–7 m^2. We set the camera to record in video mode for a duration of 10 s upon trigger. Once triggered, a camera could not be triggered again for 10 s. The camera was inspected every 3–5 months to change batteries and memory card.

All medium to large animals including mammals, terrestrial birds, and monitor lizard species captured by the camera traps were recorded and identified to species level based on Francis (2008), MacKinnon and Phillipps (1993), and Payne and Francis (2005). We classified wild boar as a morph-species (*Sus* sp.) because clear distinctions could not be made between *Sus scrofa* and *S. barbatus*, and some resembled a crossbreed between the two species. We defined a record as an independent record if the same species was not recorded during the previous 30 min. We counted the number of independent records of each species and the total length of camera-working days at each set point.

Conventionally, the trapping rate is the total number of independent records per total number of camera-working days multiplied by 100, and is termed as Relative Abundance Index (RAI) by O'Brien et al. (2003). As explained by Rowcliffe et al. (2008, 2013) and Nakashima et al. (2018), RAI can provide a index of animal density if the captured areas of a camera trap, animal speeds, and lengths of animal activity time in a day on average are not different among the plots. However, this value can be biased by the difference of camera-working days among the set points inside a plot. To offset the bias, we calculated a mean trapping rate (MTR) as an index of animal density as follows: a single-liner regression equation between the length of camera-working days at each set point and the number of independent records, with the regression line fixed to cross the point of origin. We multiplied the coefficient by 100 and defined the value as the MTR of the plot.

4.2.3   Effects of Planted Acacia Plantations

We evaluated the effects of vegetation differences caused by conversion to planted acacia forest on the trapping rate of each species in BB using the model selection method. We also compared the similarity of species composition between plots in different vegetation types using nonmetric dimensional scaling (nMDS).

To evaluate the effect of vegetation differences, we made two nested mixed Poisson models (mixed with a fixed effect and a random effect) for a number of records or number of species captured at each camera set point. The equations of the two models are as follows:

$$(\text{Model 1})\ N = wd + P + V$$

$$(\text{Model 2})\ N = wd + P,$$

where 'N' denotes the number of records, 'wd' is the length of the camera-working days, 'P' and 'V' are categorical data, representing the Plot ID and the vegetation type of the plot respectively. We compared goodness-of-fit of Model 1 and Model 2 based on Akaike's Information Criterion (AIC) values (Akaike 1973). AIC is a measure of goodness-of-fit with an added penalty for model complexity as measured by the number of fit parameters in the model (Burnham and Anderson 2002). We selected model with vegetation (model 1) as the best model when AIC of Model 1 was smaller than AIC of Model 2 and the difference was more than two.

On the other hand, nMDS is a robust unconstrained ordination method suited to community data that searches iteratively for best configuration of the plots from the dissimilarities (Gotelli and Ellison 2004). We plotted the ten plots following Bray-Curtis dissimilarity of species composition (the record numbers of each species) to investigate effect of the vegetation differences.

The generation of random set points for the cameras, and all statistical analysis were conducted using the statistical software R 3.0.1. (R Development Core Team 2013) with the packages of spatstat, lme4, and vegan.

4.3   Results

4.3.1   Faunal Composition in the Peat Swamp Forest of BB

Total camera-working days were 3978 days at 20 camera setting points in the four plots in the natural forest inside the Wildlife Reserves; 3336 days at 15 points in the three plots in the natural forest inside the protected areas of the HTIs; and 3675 days at 15 points in the three plots in the planted acacia forest inside the HTIs.

During the 10,989 camera-working days at 50 points in total, we obtained 1856 records of 19 species of mammals, 3 terrestrial birds, and 1 monitor lizard (Table 4.2). These included 11 vulnerable or endangered species on the IUCN Red

Table 4.2 Mean trapping rate of each species

Species name	Threatened status	Main food habit	Natural forest Wildlife Reserve No of records	Mean trapping rate Mean	Range	Protected area No of records	Mean trapping rate Mean	Range	Planted acacia forest No of records	Mean trapping rate Mean	Range	Difference of MTR WR-PA	WR-AF	PA-AF
No of plots			4			3			3					
Total camera working days			3978			3336			3675					
Mammals														
Echinosorex gymnura		in	9	0.18	(0.00–0.32)	10	0.28	(0.00–0.54)					a	a
Manis javanica	EN	in	3	0.08	(0.00–0.18)	3	0.13	(0.00–0.39)					a	
Nycticebus coucang	VU	he				1	0.03	(0.00–0.09)						
Presbytis femoralis percura		he & fr				3	0.11	(0.00–0.34)						
Macaca nemestrina	VU	om	196	4.81	(1.65–6.83)	109	3.22	(2.99–3.60)	36	0.96	(0.00–2.17)		a	a
Helarctos malayanus	VU	om	10	0.29	(0.00–0.50)	8	0.26	(0.20–0.34)	3	0.08	(0.00–0.23)			
Mustela flavigula		ca	1	0.02	(0.00–0.10)	1	0.02	(0.00–0.07)						
Viverra tangalunga		om	2	0.05	(0.00–0.13)	5	0.16	(0.13–0.20)	15	0.42	(0.09–0.89)		a	
Arctictis binturong	VU	om				1	0.04	(0.00–0.11)						

Species	IUCN	Diet	n	Density	95% CI	n	Density	95% CI	n	Density	95% CI				
Arctogalidia trivirgata		om	2	0.03	(0.00–0.12)				3	0.09	(0.00–0.28)				a
Paradoxurus hermaphroditus		om											a		
Hemigalus derbyanus	VU	om	14	0.31	(0.00–0.82)	10	0.29	(0.16–0.36)				a	a		a
Prionodon linsang		ca	2	0.04	(0.00–0.10)	2	0.07	(0.00–0.20)							
Herpestes brachyurus		ca				7	0.24	(0.00–0.49)						a	a
Neofelis diardi	VU	ca	1	0.03	(0.00–0.13)										
Pardofelis marmorata	VU	ca	2	0.06	(0.00–0.13)	1	0.04	(0.00–0.13)							
Prionailurus bengalensis		ca	1	0.02	(0.00–0.06)	1	0.02	(0.00–0.07)	7	0.20	(0.06–0.32)	a	a		a
Sus barbatus & S. scrofa		om	501	10.65	(3.73–20.37)	378	12.72	(4.05–26.15)	75	1.97	(0.66–3.33)	a	a		a
Tragulus kanchil		he & fr	391	8.29	(0.89–20.64)	3	0.09	(0.00–0.20)				a	a	a	a
Birds															
Lophura erythrophthalma	VU	om	23	0.46	(0.09–0.72)	3	0.09	(0.00–0.19)				a	a	a	a
Melanoperdix niger	VU	om	2	0.05	(0.00–0.10)								a		
Gallus		om							9	0.26	(0.06–0.56)	a	a		a

(continued)

Table 4.2 (continued)

		Natural forest				Protected area				Planted acacia forest			
		Wildlife Reserve											
Reptiles													
Varanus rudicollis	ca	[a]				[a]	2	0.07	(0.00–0.11)	[a]			
			Total	Mean	Range		Total	Mean	Range		Total	Mean	Range
Total			1160	29.16	9.22–51.85		548	16.42	8.75–31.76		148	4.02	2.29–6.38
Number of species			16	9.75	9–11		18	11	11		7	5.3	4–6

The effect of vegetation difference on the trapping rate was evaluated by model selection. *EN* endangered species, *VU* vulnerable species on the IUCN Red List (IUCN 2012). Main food habitat is based on Matsubayashi et al. (2007), Smythies (1999), and Myers (2009)

ca carnivore, *hf* herbivore and frugivore, *in* insectivore, *om* omnivore

[a] ΔAIC > 2

List (IUCN 2012), such as sun bear, Sunda clouded leopard (*Neofelis diardi*), marbled cat (*Paradofelis marmorata*), and Sunda pangolin (*Manis javanica*). The images of all captured species were presented in Fujita et al. (2012).

In addition to the species we recorded by camera traps, we were informed by local people that sambar deer and Sumatran tiger (*Panthera tigris*) also inhabited the natural peat swamp forest in this area. We found the skull of a male sambar deer at an illegal bird-hunting camp in the core area. It was also reported that a resident was attacked and killed by a tiger in the transition area.

The sum of MTRs of all species ranged from 9.22–51.85 (mean: 29.16), 8.75–31.76 (mean: 16.42), and 2.29–6.38 (mean: 4.02) in the natural forest inside the Wildlife Reserves, the natural forest inside the protected areas of HTIs, and the planted acacia forest inside the HTIs respectively. Camera-trap images were dominated by three species (Table 4.2), namely wild boar (*Sus* sp.), southern pig-tailed macaques, and lesser mouse-deer (*Tragulus kanchil*). Wild boar contributed to 43.2%, 69.0%, and 50.7% of all records in the Wildlife Reserves, the protected area of HTI, and the planted acacia forest, respectively. The southern pig-tailed macaques contributed to 16.9%, 19.9%, and 24.5%, respectively. Lesser mouse-deer, which were abundant only in the Wildlife Reserve, represented 33.7% of all the images. All other species had low MTRs compared to these three dominant species.

Among the three terrestrial bird species we detected, crestless fireback (*Lophura erythrophthalma*) was the dominant species in the natural peat swamp forest, accounting for 82.1% of all the records, while red jungle fowl (*Gallus gallus*) was recorded only in the planted acacia forests.

4.3.2 Differences of Faunal Composition Between Natural Peat Swamp Forests and Planted Acacia Forests

The natural peat swamp forests had a higher species richness than the planted acacia forests (Fig. 4.2). The average number of species per plot and the range were 9.75 (9–11) in the Wildlife Reserve, 11 (11) in the protected areas of HTIs, and 5.3 (4–6) in the planted acacia forests inside the HTIs (Table 4.2). Total number of recorded species was 16 in the Wildlife Reserve and 18 in the protected areas of HTIs, and seven in the planted acacia forests inside the HTIs. The higher species richness in natural peat swamp forest was the result of the number of rare species. Among the 11 vulnerable or endangered species recorded in this study, eight were recorded in the natural forests, while the other three were recorded in both vegetation types. Of the 11 vulnerable or endangered species, none were recorded only in the acacia forest.

The best mixed models selected using AIC for MTRs of 12 species including all of the abundant species in our study area were with the vegetation variable, comprising natural peat swamp forest or acacia forest (Table 4.2). Among the 12 species, the MTRs of nine species were higher in natural forest than that of the planted acacia

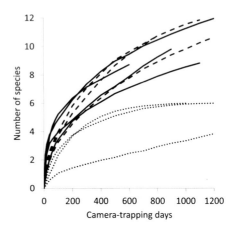

Fig. 4.2 Species accumulation curve in the ten plots. Solid line; natural forest in the Wildlife Reserve (WR1, WR2, WR3, and WR4 plots). Dash line; natural forest in the protected areas of Hutan Tanaman Industri (HTI) (PA1, PA2, and PA3 plots). Dotted line; planted acacia forest in the HTIs (AF1, AF2, and AF3 plots)

forest. On the contrary, the MTRs of Malay civet (*Viverra tangalunga*), leopard cat (*Prionailurus bengalensis*), and red jungle fowl were higher in planted acacia forest than those in the natural peat swamp forest. Besides, the best mixed model with vegetation variable between the natural peat swamp forests of the Wildlife Reserve and the protected area of HTI, was selected for only three species. Lesser mouse-deer and crestless fireback were more abundant in the Wildlife Reserve, whereas short-tailed mongoose (*Herpestes brachyurus*) was more abundant in the protected area.

The result of nMDS also showed the similarity of species composition at plots in the Wildlife Reserve and protected areas of HTI, and the dissimilarity from those at plots in the planted acacia forest (Fig. 4.3, stress: 0.084). The four plots in Wildlife Reserve and three plots in protected areas of HTI are overlapping especially on Axis 1, while three plots in the planted acacia forest were plotted apart from them.

4.4 Discussion

4.4.1 Ground-Dwelling Mammals and Bird Communities in Peat Swamp Forests

Our results indicated that the natural peat swamp forest in BB is a habitat for various mammal and terrestrial bird species including the vulnerable or endangered species. Notably, wild boar, southern pig-tailed macaques, and lesser mouse-deer were dominant species, accounting for 93% of all records (Table 4.3). One of the authors (HS) also conducted a camera-trapping survey in two lowland forests in East Kalimantan (Ratah Timber and Roda Mas) and one lowland forest in Sarawak (Anap-Muput) (Jati et al. 2018; Samejima and Hon 2019). Species compositions of the four study sites were similar, and wild boar (*Sus barbatus*), pig-tailed macaques (*Macaca nemestrina*), and mouse-deer (*Tragulus kanchil* and *T. napu*) were the top three or four most frequently recorded species also in the lowland

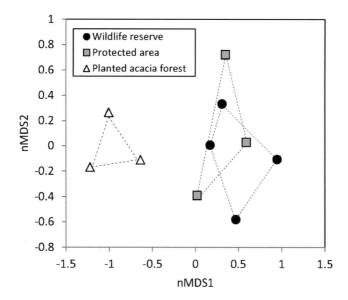

Fig. 4.3 Similarity of species composition among the ten plots shown in nonmetric dimensional scaling

Table 4.3 Proportion of records of three dominant species in natural forests in this study and three sites of lowland mixed dipterocarp forests in Sarawak and East Kalimantan

	Total camera working days	Total number of records of all species	Proportion of the records among all species		
			Sus barbatus and S. scrofa	*Macaca nemestrina*	*Tragulus kanchil and T. napu*
Natural forests in this study; Wildlife Reserve and Protected area of this (Riau)	7314	1708	52%	23%	18%
Anap-Muput (Sarawak)[a]	41,167	6924	13%	28%	1%
Ratah Timber (East Kalimantan)[b]	19,502	4160	8%	21%	23%
Roda Mas (East Kalimantan)[b]	27,426	9485	21%	18%	10%

[a]Samejima and Hon (2019)
[b]Jati et al. (2018)

forests (except *Tragulus* spp. in Anap-Muput). The proportions of wild boar found in the other sites were lower than those found in this study. While species accumulation curves in natural forests in this study were not yet saturated, species composition in peat swamp forest may be simpler than lowland rainforest in same region. Sasidhran et al. (2016) and Adila et al. (2017) also conducted a camera-trap survey in the North Selangor Peat Swamp Forest, Peninsular Malaysia, and recorded 4997 images of medium to large mammals. Among them, wild boar (*Sus scrofa*) accounted for 55% of all records. These results suggest that the natural peat swamp forest is an ecosystem dominated by wild boar. However, more studies in the natural peat swamp forest in this region are necessary to test these hypotheses.

4.4.2 Impact of Plantation Development Converted from Natural Peat Swamp Forests on Ground-Dwelling Mammals and Bird Communities

Our results indicate that the development of acacia plantations had a severe impact on species richness and abundances of ground-dwelling mammals and birds inhabiting the natural peat swamp forest. MTRs of most of the species, including the dominant species of wild boar, lesser mouse-deer and southern pig-tailed macaque, were lower in the planted acacia forests than in the natural peat swamp forests, suggesting significant population decrease of these species as a result of land-use conversion from natural peat swamp forests to industrial tree plantations. The obvious difference of species richness and the composition were also observed by Fujita et al. (2016) for bird species in same study area.

The plantation companies created substantial protected areas to maintain original vegetation. Our result showed that faunal diversity in the protected areas was as rich as in the Wildlife Reserve. However, faunal diversity inside the planted acacia forest itself was extremely low. As more than half of the species inhabiting the natural peat swamp forest were not detected in the planted acacia forests, the acacia forests are highly unsuitable as habitat for them. The planted acacia forest cannot function as a corridor for most of the ground-dwelling mammals and birds between the core area and the surrounding natural forests outside of GSK-BB.

McShea et al. (2009) also conducted a camera-trap study in an industrial tree plantation of *Acacia mangium* located in the lowland areas of central Sarawak in Borneo. In the plantation, many small patches of degraded natural forests were maintained among the planted acacia forests in addition to three large conservation zones. Presence of ground-dwelling mammal species in the remnant natural forests were higher than those in the planted acacia forests in general. Furthermore, mammal presence in the planted acacia forests were positively correlated with the closeness of the camera set points to the natural forests (McShea et al. 2009). The total number of species (n = 20) detected in the planted acacia forests was not much different (n = 24) than in the natural forests. McShea et al. concluded that planted acacia forest may not be an appropriate habitat for most endemic species, many species use the planted acacia forest to transit among patches of remnant natural forests. In short, the mosaic landscape of the acacia-natural forests may enable the persistence of many animal species in the industrial tree plantation. Nasi et al. (2009) also studied primates in an acacia plantation in lowland Sumatra, finding that natural forest fragments within industrial tree plantations do contribute to the maintenance of biodiversity. Further study may make it possible to propose appropriate design of such mosaic landscapes able to support original biodiversity as a feature of plantation development.

Some species such as Malay civet, leopard cat, and red jungle fowl may have adapted to the new plantation environment. McShea et al. (2009) also detected that wild boar (*Sus barbatus*), civet and mongoose (*Herpestes brachyurus, H. semitorquatus, Paradoxurus hemaphroditus* and *Hemigalus derbyanus*), and felidae (*Pardofelis marmorata* and *Prionailurus bengalensis*) were found in acacia forests more frequently than in logged natural forest. The adaptability of these species suggests that, whereas frugivorous species may have limited food resources in the planted acacia forest, carnivore species are not much affected.

Planted acacia forests in BB may also function as a buffer zone protecting intact natural forest in the core area from forest fires that frequently occur in the transition zone of BB (Eyes on the Forest 2013). The northern part of the transition zone in BB has burnt repeatedly and become an open area with few scattered-trees (Fig. 4.1). The plantation companies have monitored and prevented the spread of forest fires from the transition zone into their planted acacia forests.

A limited area of natural forest currently remains in lowland Sumatra. Our results indicate that natural peat swamp forest in the core area is vital to maintaining regional biodiversity, and more appropriate landscape designs, like acacia-natural forest mosaic, may boost biodiversity in industrial tree plantations.

Acknowledgments We greatly thank Mr. Canecio Munoz at Sinar Mas Forestry for his kind assistance and providing accommodation during our fieldwork. We also thank State Ministry of Research and Technology (RISTEK), the Ministry of Forestry, and Natural Resources Conservation Agency (BBKSDA), Riau, for permission to conduct this research.

We would like to express our heartfelt thanks to Alias Abdul Jalil, Yohannes Koto, Yuyu Arlan, Tju Kui Hua, Nevi Rasmika, Raffles Silaban, Zul Indra Fahmi, Edy Nazuardi, Hirimson Siahaan, Amir and other staffs from PT Sakato Pratama Makmur and PT Bukit Batu Hutani Alam, Dwi Hanum, Hutomo Rusiano, Eko, Sitinja, and Wahyu from BBKSDA, Haris Gunawan and Ridho Christina Siahaan from Riau University, Dicky from Pekanbaru, Idris, Abdul Gani, Khairin, and other field assistants from Desa Temiang, and Desa Parit I Api-Apito, who assisted in our fieldwork in Riau. We would like to also thank Yoshihiro Nakashima, who helped with discussion on camera-trapping methodology and helped to review this manuscript, and Kosuke Mizuno, who involved us in the research project in Riau, as project leader and organizer of this volume. We are also grateful to the anonymous reviewers for their constructive comments that improved our manuscript.

References

Adila N, Sasidhran S, Kamarudin N et al (2017) Effects of peat swamp logging and agricultural expansion on species richness of native mammals in Peninsular Malaysia. Basic Appl Ecol 22: 1–10. https://doi.org/10.1016/j.baae.2017.04.002

Akaike H (1973) Information theory and an extension of the maximum likelihood principle. In: Petrov BN, Csaki F (eds) Proceedings of the second international symposium on information theory. Akademiai Kiado, Budapest, pp 267–281

Anderson JAR (1961) The ecology and forest types of the peat swamp forests of Sarawak and Brunei in relation to their silviculture. Forest Department Sarawak

Anderson JAR (1976) Observations on the ecology of five peat swamp forests in Sumatra and Kalimantan. In: Peat and podzolic soils and their potential for the future. ATA Proceedings. Soil Research Institute, Bogor, pp 45–55

Bruenig EF (1990) Oligotrophic forested wetlands in Borneo. In: Goodall D (ed) Ecosystems of the world: forested wetlands. Elsevier, Amsterdam, pp 299–334

Buckley C, Nekaris KA, Husson SJ (2006) Survey of *Hylobates agilis albibarbis* in a logged peat-swamp forest: Sabangau catchment, Central Kalimantan. Primates 47:327–335. https://doi.org/10.1007/s10329-006-0195-7

Burnham KP, Anderson DR (eds) (2002) Model selection and multimodel inference: a practical information-theoretic approach. Springer, New York, p 488. https://doi.org/10.1007/b97636

Burton AC, Neilson E, Moreira D et al (2015) Wildlife camera trapping: a review and recommendations for linking surveys to ecological processes. J Appl Ecol 52:675–685. https://doi.org/10.1111/1365-2664.12432

Cannon CH, Curran LM, Marshall AJ et al (2007) Beyond mast-fruiting events: community asynchrony and individual dormancy dominate woody plant reproductive behavior across seven Bornean forest types. Curr Sci 93:1558–1566

Cheyne SM, Macdonald DW (2011) Wild felid diversity and activity patterns in Sabangau peat-swamp forest, Indonesian Borneo. Oryx 45:119–124. https://doi.org/10.1017/S003060531000133X

Cheyne SM, Thompson CJH, Phillips AC et al (2008) Density and population estimate of gibbons (*Hylobates albibarbis*) in the Sabangau catchment, Central Kalimantan, Indonesia. Primates 49: 50–56. https://doi.org/10.1007/s10329-007-0063-0

Cheyne SM, Husson SJ, Chadwick RJ et al (2010) Diversity and activity of small carnivores of the Sabangau peat-swamp forest, Indonesian Borneo. Small Carniv Conserv 43:1–7

Corlett RT (2009) The ecology of tropical East Asia. Oxford University Press, Oxford

Couwenberg J, Dommain R, Joosten H (2009) Greenhouse gas fluxes from tropical peatlands in Southeast Asia. Glob Change Biol 16:1715–1732. https://doi.org/10.1111/j.1365-2486.2009.02016.x

Ehlers Smith DA, Ehlers Smith YC (2013) Population density of red langurs in Sabangau tropical peat-swamp forest, Central Kalimantan, Indonesia. Am J Primatol 75:837–847. https://doi.org/10.1002/ajp.22145

Eyes on the Forest (2013) Sumatra's forests, their wildlife, and the climate. http://maps.eyesontheforest.or.id/. Accessed 31 Aug 2013

Felton AM, Engström LM, Felton A et al (2003) Orangutan population density, forest structure and fruit availability in hand-logged and unlogged peat swamp forests in West Kalimantan, Indonesia. Biol Conserv 114:91–101. https://doi.org/10.1016/S0006-3207(03)00013-2

Fitzherbert EB, Struebig MJ, Morel A et al (2008) How will oil palm expansion affect biodiversity? Trends Ecol Evol 23:538–545. https://doi.org/10.1016/j.tree.2008.06.012

Francis CM (2008) Field guide to the mammals of South-East Asia. New Holland Publishers, Wahroonga, p 392

Fujita MS, Irham M, Fitriana YS et al (2012) Mammals and birds in Bukit Batu area of Giam Siak Kecil - Bukit Batu Biosphere Reserve, Riau, Indonesia. Kyoto working papers on area studies 128

Fujita MS, Samejima H, Haryadi DS et al (2016) Low conservation value of converted habitat for avifauna in tropical peatland on Sumatra, Indonesia. Ecol Res 31:275–285. https://doi.org/10.1007/s11284-016-1334-2

Gaither JC Jr (1994) Understory avifauna of a Bornean peat swamp forest: is it depauperate? Wilson Bull 106:381–390

Gotelli NJ, Ellison AM (2004) A primer of ecological statistics, 1st edn. Sinauer Associates, MA, p 492

Gunawan H, Kobayashi S, Mizuno K et al (2012) Peat swamp forest types and their regeneration in Giam Siak Kecil-Bukit Batu biosphere reserve, Riau, East Sumatra, Indonesia. Mires Peat 10:5

Hooijer A, Page S, Canadell JG et al (2010) Current and future CO_2 emissions from drained peatlands in Southeast Asia. Biogeosciences 7:1505–1514

IUCN (2012) IUCN red list of threatened species. Version 2012.2. https://www.iucnredlist.org. Accessed 24 Nov 2012

Janzen DH (1974) Tropical Blackwater Rivers, animals, and mast fruiting by Dipterocarpaceae. Biotropica 6:69–103

Jati AS, Samejima H, Fujiki S et al (2018) Effects of logging on wildlife communities in certified tropical rainforests in East Kalimantan, Indonesia. For Ecol Manag 427:124–134

Johnson AE, Knott CD, Pamungkas B et al (2005) A survey of the orangutan (*Pongo pygmaeus wurmbii*) population in and around Gunung Palung National Park, West Kalimantan, Indonesia based on nest counts. Biol Conserv 121:495–507. https://doi.org/10.1016/j.biocon.2004.06.002

Kays RW, Slauson KM (2008) Remote Cameras. In: Long A, MacKay P, Zielinski WJ et al (eds) Noninvasive survey methods for carnivores. Island Press, Washington DC, pp 110–140

Kozan O (2016) Rainfall and groundwater level fluctuations in the peat swamps. In: Mizuno K, Fujita MS, Kawai S (eds) Catastrophe and regeneration in Indonesia's peatlands: ecology, economy and society. Kyoto CSEAS series on Asian studies, vol 15. NUS Press/Kyoto University Press, Singapore/Kyoto, pp 296–311

MacKinnon J, Phillipps K (1993) A field guide to the birds of Borneo, Sumatra, Java, and Bali: the Greater Sunda Islands. Oxford University Press, Oxford, p 507

Matsubayashi H, Lagan P, Majalap N et al (2007) Importance of natural licks for the mammals in Bornean inland tropical rain forests. Ecol Res 22:742–748. https://doi.org/10.1007/s11284-006-0313-4

McCallum J (2012) Changing use of camera traps in mammalian field research: habitats, taxa and study types. Mammal Rev 43:196–206

McShea WJ, Stewart C, Peterson L et al (2009) The importance of secondary forest blocks for terrestrial mammals within an *Acacia*/secondary forest matrix in Sarawak, Malaysia. Biol Conserv 142:3108–3119. https://doi.org/10.1016/j.biocon.2009.08.009

Meijaard E, Albar G, Nardiyono et al (2010) Unexpected ecological resilience in Bornean orangutans and implications for pulp and paper plantation management. PLoS One 5:e12813. https://doi.org/10.1371/journal.pone.0012813

Mohd-Azlan J (2004) Camera trapping survey in the Maludam National Park, Betong division Sarawak. Alterra/Forest Department Sarawak and Sarawak Forestry Corporation, Wageningen UR/Kuching, p 36

Morrogh-Bernard H, Husson S, Page SE et al (2003) Population status of the Bornean orangutan (*Pongo pygmaeus*) in the Sebangau peat swamp forest, Central Kalimantan, Indonesia. Biol Conserv 110:141–152. https://doi.org/10.1016/S0006-3207(02)00186-6

Myers S (2009) A field guide to the birds of Borneo. Talisman Publishing, Singapore

Nakashima Y, Fukasawa K, Samejima H (2018) Estimating animal density without individual recognition using information derivable exclusively from camera traps. J Appl Ecol 55:735–744. https://doi.org/10.1111/1365-2664.13059

Nasi R, Koponen P, Poulsen JG et al (2009) Impact of landscape and corridor design on primates in a large-scale industrial tropical plantation landscape. In: Brockerhoff EG, Jactel H, Parrotta JA et al (eds) Plantation forests and biodiversity: oxymoron or opportunity? Topics in biodiversity and conservation, vol 9. Springer, Dordrecht, pp 181–202. https://doi.org/10.1007/978-90-481-2807-5_10

National Climatic Data Center (2012) http://www7.ncdc.noaa.gov/CDO/cdosubqueryrouter.cmd. Accessed 31 Dec 2012

O'Brien TG, Kinnaird MF, Wibisono HT (2003) Crouching tigers, hidden prey: Sumatran tiger and prey populations in a tropical forest landscape. Anim Conserv 6:131–139. https://doi.org/10.1017/S1367943003003172

Page SE, Rieley JO, Doody K et al (1997) Biodiversity of tropical peat swamp forest: a case study of animal diversity in the Sungai Sebangau catchment of Central Kalimantan, Indonesia. In: Rieley JO, Page SE (eds) Biodiversity and sustainability of tropical peatlands. Samara Publishing, Cardigan, pp 231–242

Page SE, Rieley JO, Shotyk OW et al (1999) Interdependence of peat and vegetation in a tropical peat swamp forest. Philos Trans R Soc Lond B 354:1885–1897. https://doi.org/10.1098/rstb. 1999.0529

Page SE, Wűst RAJ, Weiss D et al (2004) A record of late Pleistocene and Holocene carbon accumulation and climate change from an equatorial peat bog (Kalimantan, Indonesia): implications for past, present and future carbon dynamics. J Quat Sci 19:625–635. https://doi.org/10. 1002/jqs.884

Page SE, Rieley JO, Banks CJ (2011) Global and regional importance of the tropical peatland carbon pool. Glob Change Bio 17:798–818. https://doi.org/10.1111/j.1365-2486.2010.02279.x

Partomihadjo T, Sadili A, Purwant Y (2011) Preliminary studies on structure and floristic diversity of two permanent plots in peat swamp forest biosphere Reserve Giam Siak Kecil and Bukit Batu. In: Purwanto Y, Mizuno K (eds) Proceeding of the international workshop on "sustainable management of bio-resources in tropical peat-swamp forest". The Indonesian Man and The Biosphere (MAB) UNESCO National Committee, Bengkalis, Riau, pp 47–57

Payne J, Francis CM (2005) A field guide to the mammals of Borneo, 3rd edn. The Sabah Society, Kota Kinabalu, p 332

Posa MRC (2011) Peat swamp forest avifauna of Central Kalimantan, Indonesia: effects of habitat loss and degradation. Biol Conserv 144:2548–2556. https://doi.org/10.1016/j.biocon.2011. 07.015

Posa MRC, Wijedasa LS, Corlett RT (2011) Biodiversity and conservation of tropical peat swamp forests. Bioscience 61:49–57. https://doi.org/10.1525/bio.2011.61.1.10

Quinten MC, Waltert M, Syamsuri F et al (2010) Peat swamp forest supports high primate densities on Siberut Island, Sumatra, Indonesia. Oryx 44:147–151. https://doi.org/10.1017/ S0030605309990718

R Development Core Team (2013) R: a language and environment for statistical computing. R Foundation for Statistical Computing, Vienna. http://www.R-project.org/. Accessed 1 Aug 2013

Rowcliffe JM, Carbone C (2008) Surveys using camera traps: are we looking to a brighter future? Anim Conserv 11:185–186

Rowcliffe JM, Field J, Turvey ST et al (2008) Estimating animal density using camera traps without the need for individual recognition. J Appl Ecol 45:1228–1236. https://doi.org/10.1111/j. 1365-2664.2008.01473.x

Rowcliffe JM, Kays R, Carbone C et al (2013) Clarifying assumptions behind the estimation of animal density from camera trap rates. J Wildl Manag 77:876–876. https://doi.org/10.1002/ jwmg.533

Samejima H, Hon J (2019) Diversity of medium- to large-sized ground-dwelling mammals and terrestrial birds in Sarawak. In: Ishikawa N, Soda R (eds) Anthropogenic tropical forests: human–nature interfaces on the plantation frontier. Advances in Asian human-environmental research. Springer, Singapore, pp 149–170. https://doi.org/10.1007/978-981-13-7513-2_8

Samejima H, Fujita SF, Muhammad A (2016) Biodiversity in peat swamp forest and plantations. In: Mizuno K, Fujita MS, Kawai S (eds) Catastrophe and regeneration in Indonesia's peatlands: ecology, economy and society. Kyoto CSEAS series on Asian studies, vol 15. NUS Press/Kyoto University Press, Singapore/Kyoto, pp 353–379

Sasidhran S, Adila N, Hamdan MS et al (2016) Habitat occupancy patterns and activity rate of native mammals in tropical fragmented peat swamp reserves in Peninsular Malaysia. For Ecol Manag 363:140–148. https://doi.org/10.1016/j.foreco.2015.12.037

Simbolon H, Dadi R, Haag A et al (2011) Sustainable management of tropical peat-land ecosystem in Indonesia: a resource base approach. In: Purwanto Y, Mizuno K (eds) Proceeding of the international workshop on "sustainable management of bio-resources in tropical peat-swamp forest". The Indonesian Man and The Biosphere (MAB) UNESCO National Committee, pp 1–18

Smythies BE (1999) The birds of Borneo. Natural History Publications (Borneo), Kota Kinabalu

Suzuki H, Samejima H, Fujita MS et al (2016) Outline of the survey area in Riau Province, Indonesia. In: Mizuno K, Fujita MS, Kawai S (eds) Catastrophe and regeneration in Indonesia's peatlands: ecology, economy and society. Kyoto CSEAS series on Asian studies, vol 15. NUS Press/Kyoto University Press, Singapore/Kyoto, pp 245–280

Uryu Y, Mott C, Foead N et al (2008) Deforestation, forest degradation, biodiversity loss and CO_2 emissions in Riau, Sumatra, Indonesia. WWF Indonesia Technical Report, Jakarta, p 80

van der Werf GR, Morton DC, DeFries RS et al (2009) CO_2 emissions from forest loss. Nat Geosci 2:737–738. https://doi.org/10.1038/ngeo671

Watanabe K, Masuda K, Kawai S (2016) Deforestation and the process of expansion of oil palm and acacia plantation. In: Mizuno K, Fujita MS, Kawai S (eds) Catastrophe and regeneration in Indonesia's peatlands: ecology, economy and society. Kyoto CSEAS series on Asian studies, vol 15. NUS Press/Kyoto University Press, Singapore/Kyoto, pp 281–295

Whitmore TC (1984) Tropical rain forests of the far east. Clarendon, Oxford, p 352

Whitten T, Damanik SJ, Anwar J et al (2000) The ecology of Sumatra. Periplus, Hong Kong, p 478

Yule CM (2010) Loss of biodiversity and ecosystem functioning in Indo-Malayan peat swamp forests. Biodivers Conserv 19:393–409. https://doi.org/10.1007/s10531-008-9510-5

Chapter 5
Patterns of CO_2 Emission from a Drained Peatland in Kampar Peninsula, Riau Province, Indonesia

Satyanto Krido Saptomo, Budi Indra Setiawan, Yudi Chadirin, Kazutoshi Osawa, Toshihide Nagano, Kosuke Mizuno, Dian Novarina, Susilo Sudarman, and Aulia Aruan

Abstract It was crucial to acquire soil CO_2 flux data from a bare peatland site in Kampar Peninsula, Riau Province, Indonesia so as to evaluate the carbon budget of the site in which water is managed, drained, and utilized for acacia plantation. CO_2 flux was continuously measured from July 2012 to February 2013 using an automatic soil CO_2 flux measurement system. In this study, the factors affecting carbon emission were analyzed and tested for indirect CO_2 flux estimation, and the results showed that CO_2 flux varied with weather, water, and soil-related variables, and where there was rainfall, soil temperature and soil moisture both played an important role. CO_2 flux was modeled using an artificial neural network (ANN) approach with inputs of soil moisture, temperature, and electrical conductivity (EC) as proxy variables. Based on the measurements, the total carbon dioxide (CO_2) emission during the measurement period from July 2012 to June 2013 was 52.25 t ha^{-1}. Total CO_2 emission in 2012 was estimated as 54.86 t ha^{-1} using the ANN model.

S. K. Saptomo (✉) · B. I. Setiawan · Y. Chadirin
Department of Civil and Environmental Engineering, IPB University, Bogor, Indonesia
e-mail: saptomo@apps.ipb.ac.id

K. Osawa
Utsunomiya University, Utsunomiya, Tochigi, Japan

T. Nagano
Takasaki University of Health and Welfare, Takasaki, Gunma, Japan

K. Mizuno
School of Environmental Science, University of Indonesia, Jakarta, Indonesia

Center for Southeast Asian Studies, Kyoto University, Kyoto, Japan

D. Novarina
Asia Pacific Resources International Limited, Jakarta Pusat, Indonesia

S. Sudarman
PT Riau Andalan Pulp and Paper, Pelalawan, Riau, Indonesia

A. Aruan
Alastair Fraser Forestry Foundation, Jakarta, Indonesia

© The Author(s) 2023   89
K. Mizuno et al. (eds.), *Vulnerability and Transformation of Indonesian Peatlands*,
Global Environmental Studies, https://doi.org/10.1007/978-981-99-0906-3_5

Furthermore, the results generated by the model showed that levels of CO_2 flux declined as the temperature decreased, and soil moisture increased toward soil water saturation.

Keywords MRV · Peatland · Water management · Artificial neural network

5.1 Introduction

The development of peatland often involves permanent or temporary conversion from its natural condition into bare land. Bare peat will experience exposure to natural processes in the atmosphere and on the surface, which will affect wetting and drying and, later, will lead to peat decomposition, which releases CO_2. Peat is organic soil that is composed of partially decayed vegetation and contains very little mineral content. Slow decomposition caused by anaerobic conditions is a necessary condition for its development (Belyea and Clymo 2001).

Plant remains are deposited onto the upper peat layer above the mean water table, and they experience aerobic decomposition while the remaining organic matter is buried in the saturated layer below the water table with minimal decomposition—hence, water-table depth is key to the decomposition and accumulation of peat. The lowering of the water table will remove anaerobic constraints on decomposition and stimulate the loss of peat carbon to the atmosphere (Freeman et al. 2001). A lower water table will also increase the potential for forest fires and stimulate the loss of peat carbon through combustion (Page et al. 2011; Turetsky et al. 2011).

CO_2 emission from peatland has been suspected as one of the major greenhouse gases that contribute to global warming and climate change. It is thought that this emission is dependent on the water table (Jauhiainen et al. 2008; Couwenberg et al. 2010; Jauhiainen et al. 2012; Nagano et al. 2012) and, consequently, drainage of peatland due to development and utilization will alter the emission levels of CO_2. Hooijer et al. (2006) suggest that the annual amount of CO_2 emission over peatland with a depth of 1 m of groundwater level is, on average, around 91 t ha^{-1}. However, this figure does not take into account the diurnal variation of CO_2 flux emission over the peatland, its interaction with other environmental parameters, or peat bulk density.

Net CO_2 emission can be measured using a flux tower, which requires investment in tower construction and sophisticated instruments. An indirect method using carbon budget analysis can be conducted to estimate whether the land is the source or sink of carbon emission. This requires information about carbon stored in all parts of trees and emission levels from the land. Therefore, soil emission should be continuously measured so as to acquire a CO_2 emission diurnal data series throughout the period.

This study set up a monitoring station in an area that had been developed for forest plantation operations in Kampar Peninsula. As such, the land had been drained using canals following eco-hydro management, a water management scheme implemented on site. The water table is lowered by dropping the water level in the canals that were designed and built for water management purposes. The water table is designated between 40 cm and 90 cm below the soil surface.

This study aims to quantify CO_2 emission from bare peat within the industrial forest area using continuous measurement; to determine environmental factors that affect carbon dioxide flux and develop an ANN model to estimate carbon dioxide flux with inputs of environmental factors. The measurement and analysis were conducted in the humid tropical area of Kampar Peninsula. The study focused on measuring and quantifying CO_2 flux from bare peat within the Monitoring, Reporting, and Verification (MRV) station. The quantity of CO_2 discussed in this paper is limited to CO_2 flux from bare peat (CO_2 efflux), and further use of the terms "CO_2 emission" and "CO_2 flux" will be limited to this.

5.2   Methodology

Observations were made at a specially designated MRV monitoring station on the Kampar peninsula. At this location, the land surface has been cleared, and it consists of an open area bordered by a fence to protect it from disturbances during the observation process. The water level in this concession area is maintained at an altitude of 40–90 cm below the surface. For this study, weather, soil, and CO_2 flux monitoring equipment was installed at the monitoring station.

The measurement of CO_2 flux from the soil was carried out using a closed chamber method (LICOR 8100 IRGA gas analyzer) that can measure CO_2 flux automatically based on a set time interval (Fig. 5.1). The dimension of the chamber was 20 cm in diameter and the time interval to close the chamber for data-reading was set to every 60 min (24 times per day). This was repeated three times, and the average CO_2 flux was considered to be the measurement result. At the same time, the soil temperature, electrical conductivity, and volumetric water content at a depth of 5–10 cm were measured with another sensor (5TE) and recorded by a data logger (EM50). The daily CO_2 emission was obtained by means of temporal integration

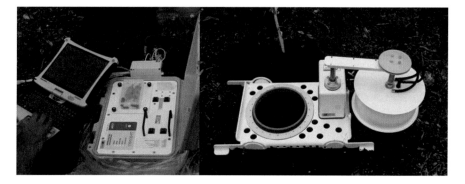

Fig. 5.1 Automatic soil CO_2 measurement system Li-8100: analyzer (left) and chamber (right)

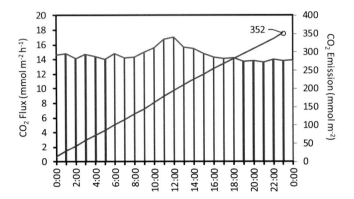

Fig. 5.2 Numerical integration to calculate daily CO_2 emission

over the CO_2 fluxes using the trapezoidal method. Then, the cumulative CO_2 emission was calculated by summing up the daily CO_2 emission for 365 days in order to obtain the total value for 1 year.

Measurement of CO_2 emission from the soil was carried out using a closed-circulation chamber methods. In this method, the soil is covered with a chamber for a designated time, and gases that come out of the soil will accumulate within the chamber. The measurement of CO_2 gas concentration was carried out by continuously flowing air from the chamber to the Infra-Red Gas Analyzer (IRGA) bench. Then, the analyzed air flowed back into the chamber. The measured change in CO_2 concentration data was then processed into CO_2 gas emitted from the ground. The whole process of measuring and processing data until the emission value was obtained was carried out automatically using the Li-8100 Automatic soil CO_2 measurement system (Fig. 5.1). The emission data obtained in this measurement was a mixture of various physical, biological, and chemical processes in the soil, resulting in CO_2 from the soil in the air per unit area.

The Li-8100 system includes a mechanical chamber and analyzer unit. The analyzer unit controls the mechanical CO_2 chamber movement and air flow using its internal pump from the chamber to the IRGA sensor, analyzes CO_2 concentration, calculates the CO_2 emission, and stores the data in a detachable memory card. The measurement was set to be conducted hourly with three replications, and the repetition data was calculated in order to get an hourly average CO_2 emission. At the end of the measurement, we produced an hourly data series for the measurement period.

The measured CO_2 concentrations during one observation event were calculated automatically by the Li-8100 system and converted into CO_2 flux with the unit of $\mu mol\ m^{-2}\ s^{-1}$. By multiplying the CO_2 flux by its molar mass divided by 10^{-6}, the CO_2 flux in units of $g\ m^{-2}\ s^{-1}$ was obtained, after which the total emission for 1 day (d) was calculated using an integration of the hourly (h) emission curve (Fig. 5.2). This integration was calculated numerically using the trapezoidal area method (Eq. 5.1) in which the daily value was the sum of the hourly values over 24 h.

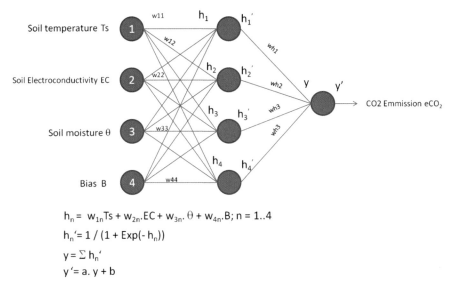

$$h_n = w_{1n}Ts + w_{2n}.EC + w_{3n}.\theta + w_{4n}.B; n = 1..4$$
$$h_n' = 1 / (1 + Exp(- h_n))$$
$$y = \Sigma h_n'$$
$$y' = a. y + b$$

Fig. 5.3 ANN model to calculate CO_2 flux

$$CO_{2\,d} = \sum_{h=1}^{24} (0.5(CO_{2\,h-1} + CO_{2\,h}).3600) \qquad (5.1)$$

Thus, the total CO_2 emission in 1 year can be calculated as follows.

$$CO_{2\,y} = \sum_{d=1}^{365} CO_{2\,d} \qquad (5.2)$$

Along with CO_2, weather and soil parameters were also measured. These measurements were part of the MRV activity conducted on the site. The measured variables were hourly rainfall, solar radiation, air temperature, air humidity, soil temperature, soil electrical conductivity, soil moisture, soil water potential, and the water table. These variables were essential to describe the environmental conditions that would influence the soilwater status and CO_2 emission from the soil. Occasional missing data of CO_2 emission was filled by the predicted value using the ANN model. Furthermore, the linear equation was used to estimate the accumulation of CO_2 emission within the missing periods.

As shown in Fig. 5.3, the ANN model consists of an input layer, a hidden layer and an output layer. The input layer had three variable inputs (soil temperature, electrical conductivity, and moisture), one bias input, and there was one variable output in the output layer, which was CO_2 flux. The model that was previously developed by Setiawan et al. (2013) in other sites had achieved considerable results with the coefficient of deterministic (R^2) being above 0.9. The related computer program (macro) was written in MS Excel, and the available Solver was used to find the optimized values of the weighted values. One reason to use these three variable inputs was knowing that there is a sensor available in the market that is capable of measuring these variables simultaneously. Once these relationships have been well proven, it is then possible to calculate CO_2 flux at the same site from these three measured variables. For other sites, similar to this one, the model needs to be calibrated through the learning or optimization process.

5.3   Results and Discussion

5.3.1 CO_2 Dynamics

Raw data of soil emission measurement is shown in Fig. 5.4, which also shows soil temperature and soil moisture at the surface layer between 5 cm and 10 cm. Judging from the varying CO_2 flux data, it can be seen that the carbon emission is influenced by environmental parameters. Figure 5.5 shows the variation of rainfall, soil moisture, and daily CO_2 emission, on site. The red line is CO_2 flux data measured with the Li-8100 system, and the green line is estimated data to fill the gap of missing data with a linear model. It is clear that CO_2 emission also varies with weather conditions, i.e., rainfall, as well as water and soil-related variables.

Carbon emission was lowered during rainfall, which increased peat moisture and lowered temperature. The CO_2 itself is produced during biological activities that are affected by moisture and temperature, and it can also be caused by human activities such as clearing and drainage of land. Thus, as soil moisture and temperature affect microbiological activity, their variation will contribute to the differences in CO_2 emission in bare peatland.

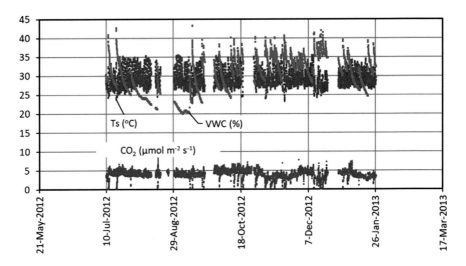

Fig. 5.4 Measured soil temperature (Ts), volumetric water content (VWC), and CO_2 flux

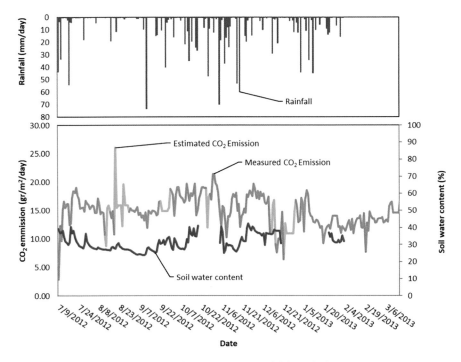

Fig. 5.5   Daily rainfall, soil volumetric water content, and CO_2 emission

5.3.2 CO_2 Correlation with Environmental Parameters

Soil CO_2 flux is the natural result of complex dynamic processes that involve organic decomposition, as well as respiration of roots and microbes (Šimůnek and Suarez 1993; Luo and Zhou 2006) in which moisture content, chemical content, and soil temperature play important roles. Berglund et al. (2010) evaluated the lysimeter method (Persson and Bergström 1991) to study the effect of temperature on CO_2 emission, which demonstrated the correlation of temperature to CO_2 by using a general model for peatland that had been drained for purposes similar to this study.

Figure 5.6 shows the individual correlations between CO_2 flux to air temperature (Ta), solar radiation (Rs), relative humidity (RH), soil temperature (Ts), soil electrical conductivity (EC), and soil volumetric water content (VWC). By looking to the low R^2, it is clear that CO_2 flux is poorly correlated with this individual variable. However, we can see that CO_2 flux tends to increase with the increase of Ta, Rs and Ts, and to decrease with the increase of RH, EC and VWC.

It is known that biological and chemical processes are dependent on temperature and energy. Water, on the other hand, has the opposite effect: Higher RH or VWC seemed to contribute to decreasing emission. It is known that CO_2 will result as oxidation occurs, i.e., the land is being drained. Conversely, oxidation will cease as the land becomes saturated with water, subsequently releasing CH_4. EC in the peat

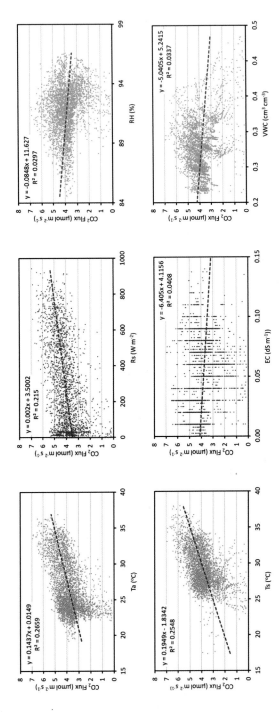

Fig. 5.6 Individual correlations between CO_2 flux with air temperature (Ta), solar radiation (Rs), relative humidity (RH), soil temperature (Ts), soil electrical conductivity (EC), and soil volumetric water content (VWC)

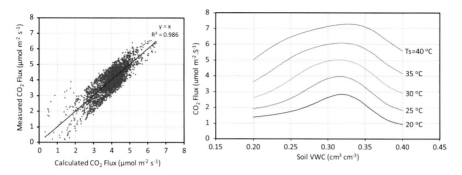

Fig. 5.7 Patterns of CO_2 flux under various soil Ts and VWC at EC = 0.04 mS cm^{-1}

can represent two parameters since the sensor actually detects conductivity, which may be due to an increase in chemical concentration or the conductive medium itself—water in the pores—and, thus, should be accounted for simultaneously with VWC. Heat and water strongly affect the emission, and two parameters that directly interact with biological processes in the soil are soil temperature and soil moisture. Therefore, these parameters can be a proxy in quantifying carbon emission.

Previously, Nagano et al. (2013) demonstrated the use of the groundwater table as a proxy parameter to estimate CO_2 emission using an empirical three-stage model (NAIS Peat Model). This is used to describe the effect of a lowering groundwater table on CO_2 release from peatlands. However, with highly porous material such as the peat in the study area and the water table at a depth of about 100 cm, capillary rise of water to the surface from groundwater is difficult to occur to wet the surface. Compared to the study area in Nagano et al. (2013), where the water table was less than 30 cm deep, it is reasonable to exclude the groundwater level from the model for the moment. In this case, CO_2 emission may not be affected by groundwater level, as it had probably been managed at a particular level in the area.

During a prolonged dry period, even though the water table was well managed at about 100 cm below the surface, the peat on the surface suffered from the very dry conditions. In a rainfall event, peat VWC immediately increased, and CO_2 flux dropped, as seen in the previous figures. These show that for the most direct estimation of CO_2, VWC can represent the water status of the peat better than the water table if VWC measurement is available.

Figure 5.7 shows the flux estimated using the ANN model with inputs of Ts, VWC, and EC versus those measured by Li-8100 at the same time. ANN parameters and constant were optimized by using the measured data. The results of the CO_2 flux calculation using the ANN model show an R^2 value of 0.986, and the model is quite accurate when used for predictions of CO_2 flux by observing proxy parameters of soil temperature, moisture, and EC. The ANN CO_2 model seems to be able to represent the characteristic of flux as dependent on soil moisture, temperature, and EC, as seen in Fig. 5.7, which is the particular example at soil EC = 0.04 mS cm^{-1}.

CO$_2$ flux increases with the increase in temperature; soil humidity, however, has a different impact on the flux. In Fig. 5.7, at any level of temperature, the flux reaches its maximum at soil moisture of between 0.31 cm^3 cm^{-3} and 0.33 cm^3 cm^{-3}. The flux decreases as soil humidity falls to a drier level or increases toward saturation. It can be concluded that either dry conditions or a saturation effect lowers emission. Dry conditions, however, will lead to soil subsidence and increase the potential for a peat fire, so wet conditions are preferable.

5.3.3   Accumulation of CO$_2$ Emission

Figure 5.8 shows the daily accumulated CO$_2$ emission measured in Kampar Peninsula from July 2012 to March 2013. The daily emission accumulated is shown by a solid rising line to the end of the available data. Measurement failures during the periods caused missing data, which were then estimated by using linear interpolation between dates with available data.

The total CO$_2$ emission throughout the measurement period is shown by the ascending line in Fig. 5.8. If this line is extended to an entire 1-year period, using this simple approach, a 1-year accumulation of CO$_2$ can be estimated at 5225 g m^{-2} (52.25 t ha^{-1}). However, this approach ignores the effect of varying soil temperature, moisture, and EC, as presented in previous section and ANN CO$_2$ model.

The ANN model for CO$_2$ emission estimation can be used when direct measurement of CO$_2$ data is not available. This can be done with a careful calibration procedure that includes direct measurements of CO$_2$ flux, which provide sufficient data. In this case, calibration had been conducted at the MRV station in Estate Meranti, and the model can be used to predict the CO$_2$ emission from the bare peatland. Figure 5.9 shows the emission series by using the ANN model, which shows that, in 2012, the total emission can be estimated as 5488 g m^{-2} (54.86 t ha^{-1}).

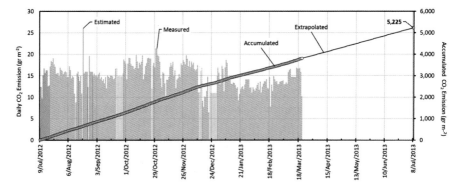

Fig. 5.8 CO$_2$ emission and accumulation

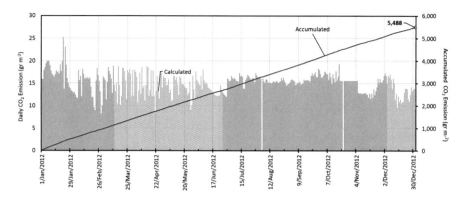

Fig. 5.9 CO_2 emission estimated using the ANN model for 2012

Long-term field data of CO_2 flux, including net ecosystem CO_2 exchange, ecosystem respiration, and soil CO_2 efflux to examine their variations for conditions of peat swamp forest in a tropical peat area in Central Kalimantan, Indonesia was presented by Hirano et al. (2016). Similar to Nagano et al. (2013), groundwater level is suggested as a practical measure to assess annual net ecosystem CO_2 exchange.

Among the results presented, the annual sum of soil CO_2 efflux or soil respiration for 2005 were 1347 g m^2 year^{-1}, 1225 g m^2 year^{-1} and 348 g m^2 year^{-1}, at the undrained, drained, and drained and burnt peat swamp forest. This is equal to 4939 g m^2 year^{-1}, 4492 g m^2 year^{-1} and 1276 g m^2 year^{-1} of CO_2, which is smaller than the estimation that has been made in this study. The data on the variation of the mean groundwater level presented at the site shows that the condition of the groundwater level presented by Hirano et al. (2016) is relatively shallower compared to the site in this study, which may lead to a lower CO_2 flux than that presented here. However, information on temperature, humidity and soil EC, and hourly measurement results are not available. Further studies on monitoring soil parameters, weather, measuring CO_2 emissions, and developing ANN models at these locations and other locations will be of interest in developing indirect CO_2 estimation methods.

5.4   Conclusion

Continuous measurement of CO_2 was conducted from July 2012 to February 2013, acquiring a CO_2 emission data series (except for missing data) during this period. Measurement showed that CO_2 flux emission varied with weather, water, and soil-related variables, especially rainfall, soil temperature, and humidity. Using linear estimation, the total estimated CO_2 emission during the measurement period from July 2012 to June 2013 is 52.25 t ha^{-1}. CO_2 emission can be estimated using the ANN approach developed in this study, with inputs of soil moisture, temperature, and EC as proxy parameters; these can be applied for rapid estimation of CO_2 emission from bare peat where it was calibrated. Total CO_2 emission in the year 2012 is estimated as 54.86 t ha^{-1} using the ANN model, which also shows that CO_2 emission fell with the decrease in temperature and increase in soil humidity toward water saturation.

Acknowledgements The authors would like to thank the Ministry of Environment and Forestry of the Republic of Indonesia for the permission given to evaluate the MRV program in the studied location from 2009 to 2014 and to express our appreciation for the support and assistance given during field surveys and throughout the program by Riau Andalan Pulp and Paper.

References

Belyea LR, Clymo RS (2001) Feedback control of the rate of peat formation. Proc Biol Sci 268(1473):1315–1321. https://doi.org/10.1098/rspb.2001.1665

Berglund Ö, Berglund K, Klemedtsson L (2010) A lysimeter study on the effect of temperature on CO_2 emission from cultivated peat soils. Geoderma 154(3–4):211–218. https://doi.org/10.1016/j.geoderma.2008.09.007

Couwenberg J, Dommain R, Joosten H (2010) Greenhouse gas fluxes from tropical peatlands in South-East Asia. Glob Change Biol 16(6):1715–1732. https://doi.org/10.1111/j.1365-2486.2009.02016.x

Freeman C, Ostle N, Kang H (2001) An enzymic 'latch' on a global carbon store. Nature 409:149. https://doi.org/10.1038/35051650

Hirano T, Sundari S, Yamada H (2016) CO_2 balance of tropical peat ecosystems. In: Osaki M, Tsuji N (eds) Tropical peatland ecosystems. Springer, Tokyo, pp 329–337. https://doi.org/10.1007/978-4-431-55681-7_21

Hooijer A, Silvius M, Wösten H et al (2006) PEAT-CO2: assessment of CO_2 emissions from drained peatlands in SE Asia. Delft Hydraulics report Q3943. Delft Hydraulics, Delft

Jauhiainen J, Limin S, Silvennoinen H et al (2008) Carbon dioxide and methane fluxes in drained tropical peat before and after hydrological restoration. Ecology 89(12):3503–3514. https://doi.org/10.1890/07-2038.1

Jauhiainen J, Hooijer A, Page SE (2012) Carbon dioxide emissions from an *acacia* plantation on peatland in Sumatra, Indonesia. Biogeosciences 9:617–630. https://doi.org/10.5194/bg-9-617-2012

Luo Y, Zhou X (2006) Soil respiration and the environment. Academic, Cambridge. https://doi.org/10.1016/B978-0-12-088782-8.X5000-1

Nagano T, Osawa K, Ishida T et al (2012) Field observation of the tropical peat soil respiration rate under various ground water levels. In: The 14th international peat congress, Stockholm, 3–8 June 2012

Nagano T, Osawa K, Ishida T et al (2013) Subsidence and soil CO_2 efflux in tropical peatland in southern Thailand under various water table and management conditions. Mires Peat 11:6

Page SE, Rieley JO, Banks CJ (2011) Global and regional importance of the tropical peatland carbon pool. Glob Change Biol 17(2):798–818. https://doi.org/10.1111/j.1365-2486.2010.02279.x

Persson L, Bergström L (1991) Drilling method for collection of undisturbed soil monoliths. Soil
    Sci Soc Am J 55(1):285–287. https://doi.org/10.2136/sssaj1991.03615995005500010050x
Setiawan BI, Saptomo SK, Bachtiar M et al (2013) Patterns of CO$_2$ flux in a bared tropical
    peatland. In: Suwardi, Nurcholis M, Agus F et al (eds) Proceeding of 11th international
    conference, the East and Southeast Asia federation of soil science societies. Land for sustaining
    food and energy security. Indonesian Society of Soil Science, Jakarta
Šimůnek J, Suarez DL (1993) Modeling of carbon dioxide transport and production in soil:
    1. Model development. Water Resour Res 29(2):487–497. https://doi.org/10.1029/92WR02225
Turetsky MR, Donahue WF, Benscoter BW (2011) Experimental drying intensifies burning and
    carbon losses in a northern peatland. Nat Commun 2:art 514. https://doi.org/10.1038/
    ncomms1523

Part II
Resilience and Adaptability of Peat Swamp Forest

Chapter 6
Termite: Friend or Foe? Conservation Values of Termites in Tropical Peat Systems

Kok-Boon Neoh, Ahmad Muhammad, Masayuki Itoh, and Osamu Kozan

Abstract Termites are the major ecosystem service providers and contribute significantly to soil processes and nutrient cycling in tropical ecosystems. The ecological services provided by termites are often discredited due to their commonly-regarded status as pest in human-dominated landscapes, however. In order to understand the potential roles of termites in peatland ecosystems, termite samplings were conducted in abandoned degraded peatland and peatland cultivated with oil palm in Riau, Sumatra. Surveys found a total of six species of termite of the family Rhinotermitidae. (*rhinotermitid*) in study plots of disturbed lands. In particular, *Coptotermes* spp. are notorious pests to oil palm, and may also be a potential pest in indigenous tree replanting programs. Based on analysis of termite feeding groups and documentation of wood susceptibility to termite attack, this study provides a reference of tree species that must be avoided in indigenous tree replanting programs so that the trophic relations of termite populations are of most benefit to peatland soil biodiversity and thereby to resilient peatland ecosystems.

Keywords Pest · Termite damage · Wood susceptibility · Ecosystem service · Oil palm

K.-B. Neoh (✉)
Department of Entomology, National Chung Hsing University, Taichung, Taiwan
e-mail: neohkokboon@nchu.edu.tw

A. Muhammad
University of Riau, Pekanbaru, Riau, Indonesia

M. Itoh
School of Human Science and Environment, University of Hyogo, Himeji, Hyogo, Japan

O. Kozan
Center for Southeast Asian Studies, Kyoto University, Kyoto, Japan

© The Author(s) 2023
K. Mizuno et al. (eds.), *Vulnerability and Transformation of Indonesian Peatlands*,
Global Environmental Studies, https://doi.org/10.1007/978-981-99-0906-3_6

6.1 Introduction

Insects contribute significantly to vital ecological functions such as pollination, decomposition, maintenance of wildlife species, and as biological control agent for crop pests (Losey and Vaughan 2006). At the same time, insects can be pests. If they appear at the wrong place or time they can threaten economic wellbeing and human health. This is particularly true when their natural habitat is disturbed by urbanization or other landscape modification, and when they are mismanaged.

Termites are major ecosystem service providers and contribute significantly in soil processes and nutrient cycling in tropical ecosystem (Holt and Lepage 2000). This vital role in ecological processes is often underacknowledged, as 5% of the 2300 termite species in the world are pests known to have negative economic impact on human settlements or agriculture (Su and Scheffrahn 2000). Recent increases in land-use intensity have elevated termites' pest status.

The Indo-Malayan region contains 62% of global tropical peatlands (Page et al. 2006). These peatlands have long served as a biodiversity hotspot for many endemic species of flora, fauna, and microbes (Yule 2010). Peatland ecosystems have also supported local community livelihoods based on long-term coexistence with peatland forests. Political policy, urbanization, and agricultural intensification, which came on the heels of increased global demand for food and fuel, resulted in the clearing of 80% of Southeast Asian peat swamp forest to make way for agro-industrial plantations (Mishra et al. 2021). In the early 2000s, 6% of tropical peatlands in the Indo-Malayan region was converted to oil-palm plantations (Koh et al. 2011), as abandoned secondary peatland increased 15% since the 1990s (Miettinen and Liew 2010). Both processes of land modification pose considerable risks to biodiversity. Even such little-regarded species as earthworms, which make similar contributions to soil processes as do termites, are disturbance-sensitive (Blanchart and Julka 1997). Owing to the unusual peat ecosystem that is highly acidic, anaerobic, and sensitive to fire and disturbed conditions, earthworms are nearly absent from disturbed peat (Cotton and Curry 1982). Termites, which tend to be much more resilient to land disturbance than are earthworms and other related species, could therefore be seen as a major soil engineer if their services could be harnessed efficiently.

As a first step toward understanding the potential roles of termites in the peat system, this study conducted termite samplings in abandoned degraded peatland and peatland cultivated with oil palm in Riau, Sumatra. The authors also combined the dataset of previous reports in disturbed and peatlands cultivated with oil palm. Based on analysis of termite feeding groups and documentation of wood susceptibility to termite attack, this study provides a reference of tree species that must be avoided in indigenous tree replanting programs so that the trophic relations of termite populations are of most benefit to peatland soil biodiversity and thereby to resilient peatland ecosystems.

6.2   Materials and Methods

An abandoned degraded peatland site and a peatland currently cultivated with 5–6 year old oil palm were selected for this study (Fig. 6.1). The sites are located in the transition zone of the GiamSiak Kecil–Bukit Batu Biosphere Reserve (0°44′–1°11′N and 0°11′–102°10′E) lying between 0 and 50 m above sea level. The study sites were 2 km apart, allowing the authors to exclude any major discrepancy in hydrological and climatic patterns, with the principal exception of canopy openness. The canopy cover in oil palm plantation registered an average of 37.8% ± 14.1 (±SE); in contrast, the abandoned degraded peatland was open shrubland.

Termite samplings were carried out in November 2012 using a standardized belt transect as described by Jones and Eggleton (2000). The belt transect comprised a survey area of 100 × 2 m and was divided into 20 sections of 5 × 2 m. A collector spent an hour on each section to collect the soldiers and worker termites from termite potential habitats such as dead tree branches, tree logs, soil under logs, termite galleries and nests. Collected termites were stored in 80% ethanol until identified. The termites were sorted to species level based on identification keys on the termite fauna in the Indo-Malayan region (Thapa 1977; Tho 1992; Gathorne-Hardy 2004).

Termite species richness was estimated using Chao 2, Incidence-based coverage estimator (ICE) and Jackknife 1, and the diversity indices (Shannon and Simpson Indices) were generated using EstimateS Version 8.2 (Colwell 2009). The relative abundance of termites was generated based on the encounter rates of termite over 20 sections. The relative abundance in abandoned land is relative to the value in oil palm cultivated land transect (relative abundance = 1).

Fig. 6.1 Abandoned degraded peatland (left) and peatland cultivated with 5–6 year-old oil palm (right) were selected as study sites. Both landscapes were dominant in transitional area of the GiamSiak Kecil-Bukit Batu Biosphere Reserve

6.3 Results

The observed species richness accumulation curve for both study sites reached an obvious asymptote close to estimated species richness curves (i.e., Chao 2, Incidence-based coverage estimator and Jackknife 1) (Fig. 6.2). In addition, all termite species were consistently encountered throughout the sampling as evidenced from no termite species only found once or twice (Singletons = 0; Doubletons = 0).

A total of six species of termite, which comprised a single family Rhinotermitidae, two subfamilies (i.e., Rhinotermitinae and Coptotermitinae) and

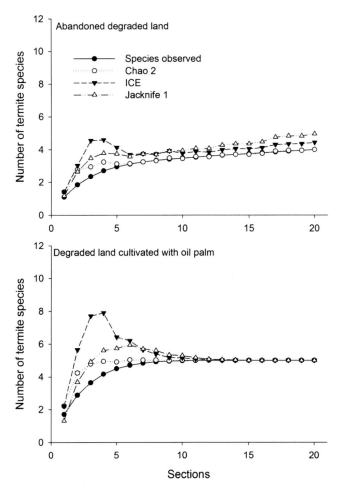

Fig. 6.2 Species accumulation curves of termites sampled in abandoned degraded peatland and peatland cultivated with oil palm in Riau, Indonesia. Species observed is the number of termites that were collected in the 100 m belt transect of 20 sections, and the numbers of termite species that were beyond observation were predicted using Chao 2, ICE and Jackknife 1

three genera (i.e., *Coptotermes*, *Schedorhinotermes*, and *Parrhinotermes*) were collected in both study sites. *Coptotermes gestroi* (Wasmann) was found only in abandoned degraded peatland, but *Parrhinotermes pygmaues* (John) and *Schedorhinotermes malaccensis* (Holmgren) were absent compared to that termite assemblage in degraded peat land cultivated with oil palm (Table 6.1).

Overall, the relative abundance of termites in abandoned degraded peatland was 65% of that in degraded peatland cultivated with oil palm (Fig. 6.3). In the abandoned degraded peat *Coptotermes kalshoveni* (Kemner), *Parrhinotermes aequalis* (Havilandi) and *Schedorhinotermes medioobscurus* (Holmgren) were predominately found (Shannon$_{exp}$ index: 3.34; Simpson index: 3.45), with relative abundances of 23%, 41%, and 32%, respectively. In contrast, the termite dominance in oil palm is even (Shannon$_{exp}$ index: 4.93; Simpson index: 5.50). Each species registered approximately 20% of encounter rate in the transect sampling (Table 6.2). All termite species found were wood feeders and wood-nesting termites.

6.4 Discussion

The termite assemblage in the transition zones of the GiamSiak Kecil-Bukit Batu Biosphere Reserve contained only a single family of lower termite–Rhinotermitidae (rhinotermitids), which consist of two subfamilies (Coptotermitinae and Rhinotermitinae), three genera, and six termite species. The result accords closely with the termite assemblage sampled in newly cleared land (Vaessen et al. 2011). The low termite-species richness in the present study sites might be the result of land disturbance during the clearing process or repeated fires, causing the collapse of susceptible termite species. Only wood-nesting termites survived at post disturbance (Neoh et al. 2016). However, this result needs to be viewed with caution, as it was based only on a single landscape. A study at a 2 year-old oil plantation conversion from mixed swamp forest in Sri Aman District, Sarawak, documented two sub-families of higher termites (i.e., Nasutitermitinae and Termitinae) and a subfamily of lower termite (Rhinotermitinae) (Vaessen et al. 2011). Another study of 5–7 year-old and 13–15 year-old oil palm plantation sites recorded more than 12 species, which comprised two subfamilies of lower termites (Coptotermitinae and Rhinotermitinae), and four subfamilies of higher termites (Nasutermitinae, Termitinae, Amitermitinae, and Macrotermitinae [only found in 15–17 year-old plot]). These findings indicate that the assemblage of termites shifts from a single dominant family to multiple families (i.e., rotten-woody feeder and soil feeder) along with the age of plantation. Kon et al. (2012) attributed this phenomenon to the improvement of soil condition across time, which favored the termite colony and territory expansion. In addition, incremental increases of dead wood, woody-plant basal area, and canopy height (as the plantation aged) may also be associated with the presence of multiple woody and soil feeders taxa in disturbed areas (Jones et al. 2003).

Termite species richness and relative abundance typically decline following logging and land-clearing activity (Jones et al. 2003). The lack of termite assemblage

Table 6.1 List of termite species found in abandoned degraded peatland and peatlands cultivated with oil palm at different ages

Taxonomic group	Feeding group[a]	Riau[b]		Sarawak[c]				Perak[d]					Pahang[e]
		ADL	5 years	CL	2 years	5–7 years	13–15 years	4 years	8 years	15 years	20 years	21 years	6–16 years
RHINOTERMITIDAE													
COPTOTERMITINAE													
Coptotermes													
C. kalshoveni	W	✓	✓										✓
C. gestroi	W	✓										✓	✓
C. curvignathus	W					✓				✓		✓	✓
C. sepangensis	W					✓	✓	✓	✓	✓	✓	✓	✓
C. borneensis	W					✓	✓	✓	✓	✓	✓		
C. travians	W												✓
RHINOTERMITINAE													
Schedorhinotermes													
S. brevialatus	W			✓	✓								✓
S. malaccensis	W		✓		✓	✓	✓						✓
S. medioobscurus	W	✓	✓		✓	✓	✓						✓
Schedorhinotermes spp.	W							✓		✓		✓	
Parrhinotermes													
P. aequalis	W	✓	✓			✓	✓						✓
P. pygmaeus	W		✓										✓
P. inaequalis	W												✓
KALOTERMITIDAE													
Kalotermitinae													
Glyptotermes brevicaudatus	W												✓

Taxonomic group	Feeding group[a]	Riau[b]		Sarawak[c]				Perak[d]					Pahang[e]
		ADL	5 years	CL	2 years	5–7 years	13–15 years	4 years	8 years	15 years	20 years	21 years	6–17 years
TERMITIDAE													
AMITERMITINAE													
Globitermes													
G. sulphurues	W											✓	
G. globosus	W						✓			✓			
Amitermes													
A. dentatus	W						✓		✓				
A. minor	W												✓
Microcerotermes havilandi	W												✓
NASUTITERMITINAE													
Bulbitermes													
B. borneesis	W				✓								
B. constrictus	W				✓								
B. germanus	W												✓
B. singaporiensis	W												✓
B. contrictiformis	W												
B. neopusillus	W												
Nasutitermes													
N. matangensiformis	W					✓	✓						
N. proatripennis	W					✓							✓
N. havilandi	W					✓	✓	✓				✓	✓
N. rectangularis	W					✓	✓						
N. johoricus	W												✓
N. longinus	W												✓

(continued)

Table 6.1 (continued)

Taxonomic group	Feeding group[a]	Riau[b] ADL	Riau[b] 5 years	Sarawak[c] CL	Sarawak[c] 2 years	Sarawak[c] 5–7 years	Sarawak[c] 13–15 years	Perak[d] 4 years	Perak[d] 8 years	Perak[d] 15 years	Perak[d] 20 years	Perak[d] 21 years	Pahang[e] 6–17 years
N. matangensis	W												✓
N. longinasoides													✓
N. neopusillus													✓
N. roboratus													✓
Hospitalitermes													
H. hospitalis	E												✓
H. umbrinus	E												✓
Oriensubulitermes													
Oriensubulitermes sp.	S			✓									
TERMITINAE													
Pericapritermes													
P. nitobei	S					✓							✓
P. paraspeciosus	S					✓	✓						✓
P. dolichocephalus	S												✓
P. latignathus	S												✓
P. mohri	S												✓
P. semarangi	S												✓
P. buitenzorgi	S												✓
Termes													
T. rostratus	W/S										✓		✓
Prohamitermes													
P. mirabilis	W/S				✓		✓						

MACROTERMITINAE		
Macrotermes		
M. gilvus	W/L	✓

ADL abandoned degraded peatland, *CL* recently cleared peatland, *W* wood feeder, *S* soil feeder, *W/S* wood/soil feeder, *W/L* wood/litter feeder, *E* microepiphyte-feeder

[a]Classification of feeding group based on Donovan et al. (2001)

[b]Present study in Bukit Batu transitional areas

[c]Data sources from Vaessen et al. (2011) and Kon et al. (2012)

[d]Data source from Cheng et al. (2008)

[e]Data source from Faszly et al. (2011) and Jalaludin et al. (2018)

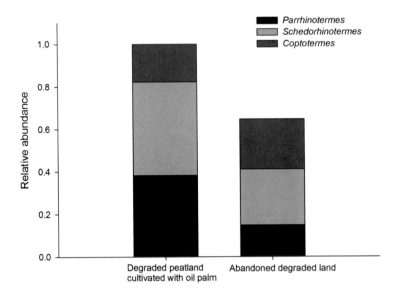

Fig. 6.3 Relative abundance of termite sampled in abandoned degraded peatland and peatland cultivated with oil palm in Riau, Indonesia. Relative abundance in abandoned land is relative to the value in oil palm cultivated land transect (relative abundance = 1)

Table 6.2 Termite species richness and diversity indices for degraded peat land and degraded peatland cultivated with oil palm

Species richness and diversity indices	Degraded peat	Degraded peat cultivated with oil palm
S_{obs}	4.00	5.00
Chao 2	4.00	5.00
ICE	4.43	5.00
Jackknife 1	4.95	5.00
Singletons	0	0
Doubletons	0	0
$Shannon_{exp}$	3.34	4.93
Simpson	3.45	5.50

in burned peatland may have reduced the functional groups of termites that are essential to soil processes. Okwakol (2000), who studied land conversion from natural forest to banana-cultivated land, has suggested that a drastic reduction of 40% of the original termite diversity potentially alters ecological processes and reduces farmland productivity over time. In the present study, termite species richness fell to one family, Rhinotermitidae. It is still difficult to estimate what rates of termite abundance and diversity can elicit significant ecosystem function in a given area. However, termite groups with proven ecosystem functionality (e.g., Macrotermitinae [Beaudrot et al. 2011], Nasutitermitinae [Jiménez et al. 2008]) were not found in our study sites. Several management practices are therefore required to maintain termite diversity when peatland is disturbed. These management techniques include: (a) using low-impact land-clearing techniques and leaving tree residues in the disturbed area (rather than burning them) in order to maintain termite populations (Bourguignon et al. 2018); and (b) maintaining patches of intact forest within a disturbed area (Neoh et al. 2017). These methods would promote the survival of termite species vulnerable to land disturbance and help restore termite biodiversity in degraded lands (Jones et al. 2003).

In mineral soils, termites are known to make subterranean tunnels that improve soil water infiltration rates, while their role in decomposing plant litter and mound building via soil translocation enriches soil nutrients and improves soil bulk density, respectively (Holt and Lepage 2000). In the Indo-Malayan region, the termite assemblage in peatlands is mainly comprised of decayed-wood feeders and soil feeders (Table 6.1). Black and Okwakol (1997) have noted that wood and soil feeders generally contain high levels of organic carbon in their nests; the latter's nest, in particular, is rich in nitrogen due to their habit of feeding on highly decaying soil-like cellulose materials. A study on soil-feeding termites in the Colombian savanna revealed that the group was reputedly more important in soil nutrient enrichment than ants (Jiménez et al. 2008). Studies on soil impact by *Nasutitermes* sp. revealed that termite mounds can act as source of nitrogen available to plants in nutrient-depleted savanna systems (López-Hernández 2001; Jiménez et al. 2006). The presence of other feeding groups in the peat system (Table 6.1) may also be vital in the optimal functioning of the ecosystem and peat restoration.

In savanna systems, grass-harvesting termites are believed to play an important role in fire management, especially when they remove grass and plant litter that could act as fuel during fire events (Davies et al. 2010). Though grass-harvesting termites are not found in the Indo-Malayan region, the presence of wood feeders of multiple taxa potentially reduces the availability of fuel materials in the peatlands.

Fire disturbance and land conversion in the peatlands under study have caused a drastic collapse in the termite assemblage, leaving only rhinotermitids behind. Although the termite assemblages reported by Vaessen et al. (2011) and Kon et al. (2012) were diverse, the relative abundance of rhinotermitids in a young oil palm plantation of 5–7 years-old was generally high compared those found of other families, as rhinotermitids account for more than 70% of the total termite encounter.

Coptotermes are highly destructive termite species in Southeast Asia. Studies conducted on oil palm growth in peatlands in peninsular Malaysia and Sarawak reported that *Coptotermes curvinagthus* (Holmgren) is a major pest that causes death in both mature and young oil palm trees (Kim Huan and Silek 2001; Cheng et al. 2008; Kon et al. 2012). Similarly, approximately 14% of rubber tree planted on peat soil were attacked by *C. curvignathus* (Indrayani et al. 2022). Rasmussen et al. (1982) suggested that an abundance of food sources from the incomplete removal of timber residue during forest clearing could account for the high prevalence of *C. curvinagthus* attack on new plantation plots. This premise has been discounted, however, as no clear relationship between the termite attack and plant residue was found in an oil palm plantation on mineral soil (Kirton et al. 1999), which could be the result of the dominance and resilience of *C. curvignathus* in newly cleared land (Cheng et al. 2008). *Coptotermes sepangensis* Krishna, *C. borneensis* Oshima, *C. gestroi*, and *C. kalshoveni* were also found in oil palm plantations, but the species generally nest in rotten wood and rarely cause tree death. Nevertheless, *C. gestroi* and *C. kalshoveni* are notorious for damaging building structures in the tropics (Kirton and Azmi 2005). Other termite species *Schedorhinotermes* spp., *Globitermes globosus* (Havilandi), and *G. sulphurues* (Havilandi) were occasional pests in oil palm plantations (Harris 1969). Their pest status may be elevated if their food sources and habitats are further disturbed.

Replanting fast-growing indigenous tree species in abandoned degraded peatland such as *Dyera lowii, Tetramerista glabra, Palaquium sumatranum, Palaquium burckii, Cratoxylon arborescens,* and *Callophyllum lowii* is favored in peat rehabilitation programs; it can also be a key element in local community activities and source of livelihood (Gunawan et al. 2011, 2013). Nevertheless, some of these tree species are susceptible to termite attack (Table 6.3). Thus we do not rule out potential termite attacks in community-planted forests, just as have occurred in oil palm plantations. To ensure the survival and growth of newly-planted trees, farmers must take into account species selection and wood resistance to termite attacks. Otherwise, sustainable pest management and preventive measures should be devised before termites become a serious threat to newly re-planted peatlands.

6.5 Conclusion

In Southeast Asia, termites are generally thought of as pests, and their ecological services to farmlands is underappreciated. With little awareness of the potentially vital role in nutrient management these soil-dwelling insects could play, farmers have often applied chemical pesticides to control termite populations in peatland oil palm plantations. Farmers drench the soil with chemicals or directly spray infestations. In most cases, these methods are repeatedly applied, as termite infestations generally reoccur after the chemical effect dissipates. Such intensive chemical interventions have wider impact on soil communities; they do not only kill pest termite species, i.e., *C. curvignathus*, but also beneficial wood- and soil-feeding termites as well as other arthropods (e.g., beetles and ants) providing natural ecosystem services. It is interesting to note that farmers in Africa possess comprehensive knowledge of termite ecology, termite species, and utilize various traditional eco-friendly control practices (Sileshi et al. 2009). Our study demonstrates that sustainable pest management and preventive measures such as tree-selection programs should be devised before termites pose a serious threat to newly planted trees.

Table 6.3 Resistance of Indonesian natural/planted timbers to termite attack

Tree species	Common name/local name	Weight loss (%)	Resistance class[a]	References
Eusideroxylon zwageri	Ulin	2.47	I	Wardani et al. (2009)
Manglietia glauca	Manglid	2.1	I	Hadjib et al. (2012)
Acacia mangium	Akasia	11.6	II	Arinana et al. (2012)
Araucaria sp.	Alau	4.52	II	Wardani et al. (2009)
Arthocarpus odoratissimus	Terap	3.82	II	Wardani et al. (2009)
Elmerrillia papuana	Wau Beech	7.27	II	Hadi et al. (2012)
Calophyllum sp.	Bintagur	4.77	II	Hadi et al. (2012)
Swintonia glauca	Sumpung	7.42	II	Wardani et al. (2009)
Pometia pinnata	Taun	6.40	II	Hadi et al. (2012)
Garcinia sp.	Garcinia	6.09	II	Hadi et al. (2012)
Melia azedarach	Mindi	10.3	III	Hadi (2014)
Peronema canescens	Sungkai	12.46	III	Wardani et al. (2009)
Canarium sp.	Canarium	7.99	III	Hadi et al. (2012)
Dyera costulata	Pantung	15.12	IV	Wardani et al. (2009)
Durio zibethinus	Durian	15.52	IV	Wardani et al. (2009)
Mangifera sp.	Mangga	14.80	IV	Wardani et al. (2009)
Anthocephalus macrophylla	Red Jabon	16.61	IV	Wardani et al. (2012)
Pinus merkusii Jungh	Pine	16.9	IV	Jasni and Rachman (2009)
Shorea acuminate	Meranti merah	20.86	IV	Wardani et al. (2009)
Elmerillia ovalis	Cempaka Hutan	19.4	V	Hadjib et al. (2012)
Elaeis guineensis	Oil palm	26.29	V	Dungani et al. (2013)
Ochroma lagopus	Balsa	37.2	V	Hadi et al. (2012)
Hevea brasuliensis	Rubber	19.40–39.89	V	Wardani et al. (2009), Hadjib et al. (2012), and Wardani et al. (2012)
Anthocephalus cadamba	White Jabon	32.40	V	Wardani et al. (2012)
Peronema sp.	Pulai	36.4	V	Hadi (2014)

(continued)

Table 6.3 (continued)

Tree species	Common name/local name	Weight loss (%)	Resistance class[a]	References
Paraserianthes falcataria	Sengon	23.10–37.3	V	Wardani et al. (2009), Arinana et al. (2012), and Hadi (2014)

[a]Wood resistance to termite attack based on the weight loss of wood consequence upon termite attack set by Standard Nasional Indonesia (SNI 2006)

Acknowledgments We thank Mr. Nur who kindly provided accommodations during our field survey. K.-B.N. was an international researcher fellow of the Japan Society for the Promotion of Science. This study was partly funded by Large-Scale Research Program "Promoting the Study of Sustainable Humanosphere in Southeast Asia" funded by the Japanese Ministry of Education, Culture, Sports, Science, and Technology (MEXT), 2011–2016, and the institutional collaboration feasibility studies "Toward the Regeneration of Tropical Peatland Societies: Building International Research Network on Paludiculture and Sustainable Peatland Management", Research Institute for Humanity and Nature (RIHN).

Referecses

Arinana, Tsunoda K, Herliyana EN et al (2012) Termite-susceptible species of wood for inclusion as a reference in Indonesian standardized laboratory testing. Insects 3(2):396–401. https://doi.org/10.3390/insects3020396

Beaudrot L, Du Y, Rahman Kassim A et al (2011) Do epigeal termite mounds increase the diversity of plant habitats in a tropical rain forest in Peninsular Malaysia? PLoS One 6(5):19777. https://doi.org/10.1371/journal.pone.0019777

Black HIJ, Okwakol MJN (1997) Agricultural intensification, soil biodiversity and agroecosystem function in the tropics: the role of termites. Appl Soil Ecol 6(1):37–53. https://doi.org/10.1016/S0929-1393(96)00153-9

Blanchart E, Julka JM (1997) Influence of forest disturbance on earthworm (Oligochaeta) communities in the Western Ghats (South India). Soil Biol Biochem 29(3–4):303–306. https://doi.org/10.1016/S0038-0717(96)00094-6

Bourguignon T, Dahlsjö CAL, Salim KA et al (2018) Termite diversity and species composition in heath forests, mixed dipterocarp forests, and pristine and selectively logged tropical peat swamp forests in Brunei. Insect Soc 65(3):439–444. https://doi.org/10.1007/s00040-018-0630-y

Cheng S, Kirton LG, Gurmit S (2008) Termite attack on oil palm grown on peat soil: identification of pest species and factors contributing to the problem. Planter 84(991):200–210

Colwell RK (2009) EstimateS 8.2.0.: statistical estimation of species richness and shared species from samples. User's guide and application. Department of Ecology and Evolutionary Biology, University of Connecticut, Storrs

Cotton DCF, Curry JP (1982) Earthworm distribution and abundance along a mineral-peat soil transect. Soil Biol Biochem 14(3):211–214. https://doi.org/10.1016/0038-0717(82)90026-8

Davies AB, Parr CL, Van Rensburg BJ (2010) Termites and fire: current understanding and future research directions for improved savanna conservation. Austral Ecol 35(4):482–486. https://doi.org/10.1111/j.1442-9993.2010.02124.x

Donovan SE, Eggleton P, Bignell DE (2001) Gut content analysis and a new feeding group classification of termites. Ecol Entomol 26(4):356–366. https://doi.org/10.1046/j.1365-2311.2001.00342.x

Dungani R, Islam MN, Abdul Khalil HPS et al (2013) Termite resistance study of oil palm trunk lumber (OPTL) impregnated with oil palm shell meal and phenol-formaldehyde resin. Bioresources 8(4):4937–4950

Faszly R, Mohammad-Faris ME, Azizil-Alimin MM et al (2011) Termites of oil palm on peat soil: a 10-year collection from Endau Rompin. Serangga 16(2):37–56

Gathorne-Hardy FJ (2004) The termites of Sundaland: a taxonomic review. Sarawak Museum J 60(81):89–133

Gunawan H, Kobayashi S, Mizuno K et al (2011) Progress on restoration experiments of degraded peat swamp forest ecosystem in the Giam Siak Kecil-Bukit Batu Biosphere Reserve, Riau, Indonesia. In: Purwanto Y, Mizuno K (eds) Proceeding of the international workshop on sustainable management of bio-resources in tropical peat-swamp forest, Cibinong, 2011

Gunawan H, Kobayashi S, Mizuno K et al (2013) Sustainable rehabilitation of tropical peat swamp forest ecosystem in Giam Siak Biosphere Reserve, Riau, Indonesia: an integrated approach. In: Ishimaru K, Kobayashi S (eds) Proceedings of international workshop on incentive of local community for REDD and semi-domestication of non-timber forest products, Kyoto, 2011

Hadi YS (2014) Feeding rate as a consideration factor for successful termite wood preference tests. In: Forschler BT (ed) Proceedings of the 10th pacific termite research group conference, Kuala Lumpur, 2014

Hadi YS, Massijaya MY, Hadjib N et al (2012) The resistance of six Papua New Guinea woods to subterranean termite attack. In: Proceedings of the 9th Pacific-Rim termite research group conference, Honoi, February 2012. Science and Technics Publishing House, Hanoi, pp 91–94

Hadjib N, Massijaya MY, Hadi YS et al (2012) Resistance of three small diameter logs to subterranean termite attack. In: Proceedings of the 9th Pacific-Rim termite research group conference, Honoi, February 2012. Science and Technics Publishing House, Hanoi, pp 95–98

Harris WV (1969) Termites as pests of crops and trees. Commonwealth Agricultural Bureau, London

Holt JA, Lepage M (2000) Termites and soil properties. In: Abe T, Bignell DE, Higashi M (eds) Termites: evolution, sociality, symbioses, ecology. Kluwer Academic Publishers, Dordrecht, pp 389–407. https://doi.org/10.1007/978-94-017-3223-9_18

Indrayani Y, Setyawati D, Maryani Y et al (2022) A termite attack on rubber plantation on peat soil: level of damage and identification of pest species. In: 2nd international conference on tropical wetland biodiversity and conservation, Banjarbaru City, October 2021. IOP conference series: earth and environmental science, vol 976. IOP Publishing, Bristol, 012006. https://doi.org/10.1088/1755-1315/976/1/012006

Jalaludin NA, Rahim F, Yaakop S (2018) Termite associated to oil palm stands in three types of soils in Ladang Endau Rompin, Pahang, Malaysia. Sains Malays 47(9):1961–1967. https://doi.org/10.17576/jsm-2018-4709-03

Jasni HR, Rachman O (2009) The resistance of pine wood from timber estate against termite at various levels of tree age. In: Proceedings of the 6th conference of the Pacific Rim termite research group, Kyoto, 2–3 March 2009

Jiménez JJ, Decaëns T, Lavelle P (2006) Nutrient spatial variability in biogenic structures of Nasutitermes (Termitinae; Isoptera) in a gallery forest of the Colombian 'Llanos'. Soil Biol Biochem 38(5):1132–1138. https://doi.org/10.1016/j.soilbio.2005.09.026

Jiménez JJ, Decaëns T, Lavelle P (2008) C and N concentrations in biogenic structures of a soil-feeding termite and a fungus-growing ant in the Colombian savannas. Appl Soil Ecol 40(1):120–128. https://doi.org/10.1016/j.apsoil.2008.03.009

Jones DT, Eggleton P (2000) Sampling termite assemblages in tropical forests: testing a rapid biodiversity assessment protocol. J Appl Ecol 37(1):191–203. https://doi.org/10.1046/j.1365-2664.2000.00464.x

Jones DT, Susilo FX, Bignell DE et al (2003) Termite assemblage collapse along a land-use intensification gradient in lowland Central Sumatra, Indonesia. J Appl Ecol 40(2):380–391. https://doi.org/10.1046/j.1365-2664.2003.00794.x

Kim Huan L, Silek B (2001) Termite infestation on oil palms planted on deep peat in Sarawak: tradewinds experience. In: Cutting-edge technologies for sustained competitiveness. Proceedings of the 2001 PIPOC international palm oil congress, Kuala Lumpur, August 2001. Malaysian Palm Oil Board, Selangor, pp 355–368

Kirton LG, Azmi M (2005) Patterns in the relative incidence of subterranean termite species infesting buildings in Peninsular Malaysia. Sociobiology 46(1):1–15

Kirton LG, Brown VK, Azmi M (1999) Do forest-floor wood residues in plantations increase the incidence of termite attack?—testing current theory. J Trop For Sci 11:218–239

Koh LP, Miettinen J, Liew SC et al (2011) Remotely sensed evidence of tropical peatland conversion to oil palm. PNAS 108(12):5127–5132. https://doi.org/10.1073/pnas.1018776108

Kon TW, Bong CF, King JH et al (2012) Biodiversity of termite (Insecta: Isoptera) in tropical peat land cultivated with oil palms. Pak J Biol Sci 15(3):108–120. https://doi.org/10.3923/pjbs.2012.108.120

López-Hernández D (2001) Nutrient dynamics (C, N and P) in termite mounds of *Nasutitermes ephratae* from savannas of the Orinoco Llanos (Venezuela). Soil Biol Biochem 33(6):747–753. https://doi.org/10.1016/S0038-0717(00)00220-0

Losey JE, Vaughan M (2006) The economic value of ecological services provided by insects. Bioscience 56(4):311–323. https://doi.org/10.1641/0006-3568(2006)56[311:TEVOES]2.0.CO;2

Miettinen J, Liew SC (2010) Degradation and development of peatlands in Peninsular Malaysia and in the islands of Sumatra and Borneo since 1990. Land Degrad Dev 21(3):285–296. https://doi.org/10.1002/ldr.976

Mishra S, Page SE, Cobb AR et al (2021) Degradation of Southeast Asian tropical peatlands and integrated strategies for their better management and restoration. J Appl Ecol 58(7):1370–1387. https://doi.org/10.1111/1365-2664.13905

Neoh KB, Bong LJ, Muhammad A et al (2016) The impact of tropical peat fire on termite assemblage in Sumatra, Indonesia: reduced complexity of community structure and survival strategies. Environ Entomol 45(5):1170–1177. https://doi.org/10.1093/ee/nvw116

Neoh KB, Bong LJ, Muhammad A et al (2017) The effect of remnant forest on insect successional response in tropical fire-impacted peatland: a bi-taxa comparison. PLoS One 12(3):e0174388. https://doi.org/10.1371/journal.pone.0174388

Okwakol MJN (2000) Changes in termite (Isoptera) communities due to the clearance and cultivation of tropical forest in Uganda. Afr J Ecol 38(1):1–7. https://doi.org/10.1046/j.1365-2028.2000.00189.x

Page SE, Rieley JO, Wüst R (2006) Lowland tropical peatlands of Southeast Asia. In: Martini IP, Martínez Cortizas A, Chesworth W (eds) Peatlands: evolution and records of environmental and climate changes, Developments in earth surface processes, vol 9. Elsevier, Amsterdam, pp 145–172. https://doi.org/10.1016/S0928-2025(06)09007-9

Rasmussen AN, Kanapathy K, Santa Maria N et al (1982) Establishment of oil palm on deep peat from jungle. In: Pushparajah E, Chew PS (eds) The oil palm in agricultural in the eighties. The Incorporated Society of Planters, Kuala Lumpur, pp 641–651

Sileshi GW, Nyeko P, Nkunika POY et al (2009) Integrating ethno-ecological and scientific knowledge of termites for sustainable termite management and human welfare in Africa. Ecol Soc 14(1):48

SNI (Standar Nasional Indonesia) (2006) SNI 01.7207-2006: Uji ketahanan kayu dan produk kayu terhadap organisme perusak kayu. Badan Standardisasi Nasional, Jakarta

Su NY, Scheffrahn RH (2000) Termites as pest of buildings. In: Abe T, Bignell DE, Higashi M (eds) Termites: evolution, sociality, symbioses, ecology. Kluwer Academic Publishers, Dordrecht, pp 437–453. https://doi.org/10.1007/978-94-017-3223-9_20

Thapa RS (1977) Termites of Sabah. Sabah Forestry Department, Sandakan

Tho YP (1992) Termites of peninsular Malaysia. Forest Research Institute of Malaysia, Kuala Lumpur

Vaessen T, Verwer C, Demies M et al (2011) Comparison of termite assemblages along a landuse gradient on peat areas in Sarawak, Malaysia. J Trop For Sci 23(2):196–203

Wardani L, Subaru D, Jasni et al (2009) Termite resistance of some woods from natural and plantation forests in South Kalimantan Indonesia. In: Proceedings of the 6th Pacific Rim termite research group conference, Kyoto, 2–3 March 2009

Wardani L, Risnasari I, Yasni Hadi YS et al (2012) Resistance of Jabon timber modified with styrene and MMA against soil termites and dry wood termites. In: Proceedings of the 9th Pacific- Rim Termite Research Group Conference, Science and Technics Publishing House, Hanoi, pp 73–78

Yule CM (2010) Loss of biodiversity and ecosystem functioning in Indo-Malayan peat swamp forests. Biodivers Conserv 19(2):393–409. https://doi.org/10.1007/s10531-008-9510-5

Chapter 7
The Timber Processing and Retail Sectors in Pekanbaru, Riau: Toward Reforestation by Local People

Haruka Suzuki

Abstract The integral role of the timber industry in sustaining local reforestation efforts is often overlooked. The processing and retail sectors of the timber industry are significant in that they add value to forest resources, providing local jobs and lumber supplies. This study examines these sectors in Riau, Indonesia, where deforestation and degraded peatlands have been accelerating. Based on a field survey of randomly selected molding mills and timber kiosks in Pekanbaru, the author examines the nature and scale of timber production, sales, and supply chains. Both the molding mills and timber kiosks could be classified into three types: those that function as subcontractors of large pulp and paper companies, those that operate independently and provide timber supplies for local construction needs, and those that specialize in higher-end wood processing. Those providing local timber supplies demonstrated more adaptability in terms of meeting market demands. Depending on the level of the operation's independence, timber supply chains were more diversified in terms of the relationships among loggers, distributors, and retailers. To develop the timber sector in the context of local reforestation, it is necessary to consider how local people connect with the local timber processing and retail sectors and how to improve local timber supply chains. Reforestation programs must construct sustainable management systems that restore forests while at the same time using forest resources. This includes developing land ownership and use systems to produce timber sustainably and replanting tree species that are useful as timber.

Keywords Tropical timber · Timber supply · Reforestation · Forest management · Riau · Indonesia

H. Suzuki (✉)
Faculty of Humanities and Sciences, Kobe Gakuin University, Kobe, Japan
e-mail: h-suzuki@human.kobegakuin.ac.jp

© The Author(s) 2023
K. Mizuno et al. (eds.), *Vulnerability and Transformation of Indonesian Peatlands*, Global Environmental Studies, https://doi.org/10.1007/978-981-99-0906-3_7

7.1 Introduction

The eastern province of Riau in Indonesia has been experiencing an accelerated rate of deforestation and peatland degradation since the 1990s. Furthermore, since the early 2010s, economic and population growth has precipitated a rise in the demand for timber for building construction. Local reforestation efforts are expanding in many areas in Riau, particularly in the context of rehabilitating degraded peatlands. At the same time, the Indonesian government has been developing several reforestation programs with the aim to increase both local timber supplies and employment opportunities. For example, in parts of Java, the planting of *Paraserianthes falcataria* and *Albizia chinensis* has successfully benefited the local wood processing industry (van Noordwijk et al. 2007).

Timber supply networks in Java directly connect local timber producers and wood processing companies, stabilizing the timber supply and decreasing gaps in timber distribution (Iwanaga and Masuda 2012, pp. 42–47). Although numerous studies examine the question of "illegality" in the production of timber in Indonesia, particularly in the context of political and economic change in the country (see Obidzinski 2005, pp. 193–205; Casson and Obidzinski 2002, pp. 2133–2151), the question of how timber production and supply by local communities could support reforestation requires more investigation and discussion. This chapter examines the socioeconomic characteristics of the timber processing and retail sectors—which are important hubs of the local timber supply system that connect timber producers, suppliers, and consumers—and discusses these sectors in the context of reforestation by local communities.

The chapter focuses on Pekanbaru, the capital city of Riau. Although the use of concrete has increased in building construction in many areas of Indonesia, timber remains an essential building material. Therefore, timber demand is expected to continue to increase in Riau due to sustained economic and population growth (BPS Provinsi Riau 2019, 2022a, b, c, d). Pulp and paper production in Riau is also expected to increase, as Riau is a main area of pulp production in Indonesia and pulp production has rapidly increased in recent years (Kementerian Lingkungan Hidup dan Kehutanan 2019, 2020). How, then, can Riau fill these demands in an environmentally and economically viable way?

Timber has long been one of eastern Sumatra's primary marketable products and subsequently is a valuable commodity. Historical research on the timber industry in eastern Sumatra has paid attention to the socioeconomic aspects of logging and timber trading. For example, Barnard (1998) studied the timber trade in premodern Siak, a kingdom located on the east coast of Sumatra, from 1723 to 1946. Local communities living along the Siak River—which is the deepest waterway in the Malay Archipelago and thus an important conduit for trade in the region—exported timber, for which Siak was well known throughout the Straits of Malacca and as far away as Java (Barnard 1998). Pastor (1927) examined the logging practices of the ethnic Chinese in the coastal area of east Sumatra during the 1860s and 1870s. They formed groups of 20–30 people and logged while staying in small sheds in the forest.

The timber was then exported to Singapore. By the nineteenth century, the importance of the Siak timber trade was cemented, as other regions of Southeast Asia, particularly those with easily accessible teak forests, became deforested (Barnard 1998). Indeed, ethnic Chinese logging groups continued to actively operate until 1962, when trade between eastern Sumatra and Singapore was controlled by regulation.

After the Suharto period, economic and political crises caused political power to shift from centralized, authoritarian rule to democratization and decentralization, impacting local timber business in ways that are discussed in the context of political ecology. In analyzing the networks of exchange and accommodation surrounding illegal logging, McCarthy (2006) found that in some local villages, people engage in illegal logging when the price of cash crops, such as *nilam* (patchouli oil), decreases and they need other sources of income. Yet while such practices offer short-term profit, illegal logging is understood to be inherently unsustainable, because the main income source, commodity production, remains only partially developed (Potter and Badcock 2004, pp. 341–356). Iwanaga et al. (2014) investigated how local timber production functions in the global carbon credit system, using the Melaleuca (*Melaleuca cajupti*) market in Indonesia as a case study of forest management and timber production in a REDD+ (reducing emissions from deforestation and forest degradation) project. Finding that demand for medium-sized Melaleuca timber increased, Iwanaga concludes that Melaleuca silviculture and timber production could be one option to increase local people's income in the REDD+ project (Iwanaga et al. 2014).

Previous studies such as these illustrate the integral role that the timber industry plays in sustaining local livelihoods and forests. To improve understanding of this role, this chapter clarifies the socioeconomic characteristics of local timber processing and retail operations, focusing on molding mills and timber kiosks in Pekanbaru, Riau. Molding mills are outfits that have a number of circular and band saws to process logs and lumber, but they do not operate sawmills, which process raw wood into logs. Timber kiosks simply sell processed timber and do not have any machines to process the timber.

In describing these operations, I also briefly explain the relationship between pulp and paper companies, which produce and process timber on a large scale, and local business, which produce and retail timber on a small scale. Finally, I conclude by noting some challenges and opportunities of the local timber industry vis-à-vis reforestation efforts by local people.

7.2   Method

7.2.1   Research Site

Research was conducted in Pekanbaru, Riau, Indonesia (see Figs. 7.1 and 7.2). Although very few logs are produced in Pekanbaru, sawn timber and plywood are

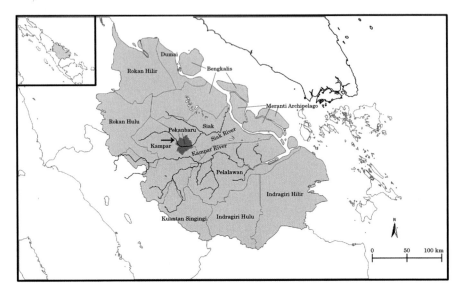

Fig. 7.1 Map of Riau Province, Indonesia

produced in the city, making it a timber processing and retail area of Riau. In addition, Riau is the center of the pulp and paper industry in Indonesia, with several major companies operating in the province.

Until several years ago, local molding mills and kiosks in Pekanbaru were mainly located in two areas. The first area, along the Siak River, was the main access point for transporting logs when logging was active along its tributaries.

The second area is in the southern part of Pekanbaru, where molding mills and timber kiosks line the major roads for local commodity distribution in Riau. These local timber businesses continue to be active, for example on the north-south road to Dumai, a mid-coastal city and the east-west road stretching from Tembilahan in Indragiri Hilir to Rengat in Indragiri Hulu.

7.2.2   Field Survey and Interviews with Stakeholders

Official records and statistical data do not accurately reflect the timber operations in Pekanbaru, because some businesses engage in illegal practices as defined by current forestry policy.[1] To address this gap in knowledge, a field survey was conducted in November 2011 and February 2012 of randomly selected molding mills and timber

---

[1] This is based on an interview with the officer in the Department of Forest Production Utilization and Monitoring (*Balai Pemantauan dan Pemanfaatan Hutan Produksi*, BP2HP) in Pekanbaru, Riau, on October 28, 2011.

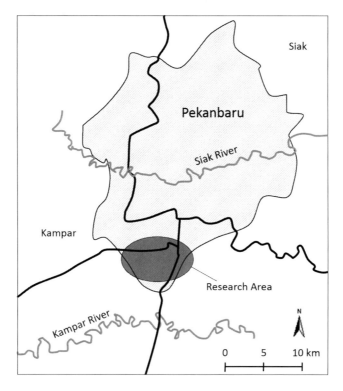

Fig. 7.2 Research area and major roads in Pekanbaru, Riau, Indonesia. The black lines are the roads

kiosks along the major roads in Pekanbaru. The author also interviewed the timber business managers and workers about their production scale, the timber species and sizes that they process and sell, their management practices, and supply chains, with a focus on the relationships among loggers, distributors, and retailers. The interviews provide useful information and perspectives for better understanding the local timber processing and retail sectors in Pekanbaru.

7.3 Findings

7.3.1 Local Molding Mills and Timber Kiosks

7.3.1.1 Business Characteristics and Scale of Production and Sales

The molding mills are classified into three types based on their business characteristics, as shown in Table 7.1. The first type represents the majority of those in Pekanbaru and includes those molding mills that functioned as subcontractors of

Table 7.1 Surveyed molding mills according to business characteristics and production scale (Source of data: Author's field survey)

No.	Molding mill type according to business characteristics	Number of molding mills by amount of timber retailed (m³/month)								Total
		0 m³	<50 m³	<100 m³	<150 m³	<200 m³	<250 m³	<300 m³		
1	Processing wooden pallets to sell to the pulp companies	0	0	0	2	2	1	1		6
2	Processing timber for house construction	0	0	1	1	1	0	0		3
3	Processing timber for doors, window frames, and roof ornaments	0	1	0	0	0	0	0		1
Total		0	1	1	3	3	1	1		10

large pulp and paper companies. They processed wooden pallets for the pulp and paper companies, which then used the pallets to export pulp to foreign countries. This type of molding mill had a comparatively large production volume compared to the other types, which processed timber for building construction and/or furniture. Many of these molding mills began pallet processing a few years prior to the survey. Several managers of these molding mills noted that one of the reasons they work with the pulp and paper companies is that it does not require them to obtain any permission to process and retail timber, because they can work under the permission of the larger companies. Asia Pulp & Paper Co. Ltd. (APP) and Asia Pacific Resources International Holding Limited (APRIL), major pulp and paper companies in Indonesia, plant and harvest Acacia in plantations and process the cut wood into pulp at their plants. Other than the pallets, the local molding mills and timber kiosks mainly sell timber, such as building materials, to the local communities. This division of labor in the supply chain eliminates competition between the pulp companies and local operations. However, some Type 1 mills develop a dependence on the pulp and paper companies.

The second type of molding mill refers to those that processed timber to sell to both local people and government entities for building construction. This included sales of timber to the local government, which had a number of ongoing infrastructure and public projects that required timber. The production scale of these mills depended on the scale of government projects.

The third type includes the molding mills that processed timber specifically for doors, window frames, and roof ornamentation. These mills produced less than 50 m^3 of processed timber per month.

The business characteristics of the surveyed timber kiosks are classified into three categories in Table 7.2. The first type consists of timber kiosks that retail timber to the mills that are subcontracted by pulp and paper companies, local people, and local government. This type of kiosk sold more species and sizes of timber than the other two types of kiosks. Moreover, overall sales were higher in this kiosk than in the other types, with the amounts of timber product sold ranging from 50 m^3 to 200 m^3 per month.

The second type includes timber kiosks that retailed timber to local people and government for house and building construction. Their sales amounted to less than 150 m^3 per month. The third category consists of kiosks that retailed timber for doors, window frames, and roof ornaments. Their sales amounted to less than 50 m^3 per month.

7.3.1.2   Managers' Ethnicities and Hometowns

Based on interviews with managers and workers, almost all the managers of both the molding mills and timber kiosks surveyed were born in Riau. Most managers were from Pekanbaru, but some were from Kampar and Siak, neighboring regencies to Pekanbaru. They were predominantly Melayu people, who are traditionally distributed along the east coast of Sumatra. Some Batak and Minangkabau managers were

Table 7.2 Surveyed timber kiosks according to business characteristics and sales volume (Source of data: Author's field survey)

No.	Timber kiosk type according to business characteristics	Number of timber kiosks by amount of timber retailed (m³/month)							
		0 m³	<50 m³	<100 m³	<150 m³	<200 m³	<250 m³	<300 m³	Total
1	Retailing timber to subcontract mills of the pulp company, local people, and local government	0	0	1	3	2	0	0	6
2	Retailing timber to local people and local government	0	3	3	2	0	0	0	8
3	Retailing timber to kiosks which process and sell doors, window frames, and roof ornaments	1	0	0	0	0	0	0	1
Total		1	3	4	5	2	0	0	15

born in North Sumatra and West Sumatra, but had been living in Pekanbaru for more than 10 years at the time of the study.

7.3.1.3 Workers

Both the molding mills and timber kiosks were mainly family-operated businesses. As shown in Table 7.3, on average, two family members, for example, the manager's wife and his relative, worked in the molding mills and timber kiosks. In addition to this core management team, migrant workers were also common in the molding mills and timber kiosks. Based on interviews with managers, working in a molding mill or timber kiosk was a typical way for migrants to make a living in Riau. Almost all these workers were young men in their 20s and 30s, who came from North Sumatra, West Sumatra, and Java.

7.3.1.4 Business Management

It was extremely difficult to obtain a clear picture of the business operations of molding mills and timber kiosks due to the persistent sensitivity around illegal logging. Many managers did not want to reveal details about their businesses. One molding mill, however, did share about its operations in an interview on October 22, 2011. The monthly income and expenses of this molding mill in October 2011 are presented in Table 7.4.

The Type 1 mill processed wooden pulp pallets for pulp and paper companies. It was managed by the owner; his wife and uncle assisted in management (their salaries are not included in Table 7.4). The owner employed 12 workers. Labor costs amounted to IDR 1,415,000 per month in total. This value exceeds IDR 1,200,000 (approximately USD 80), which is the average monthly wage for a timber processing worker in Indonesia (BPS 2012). He also had to pay annual land rental costs, as the business was operated on borrowed land.

Table 7.3 Background of workers in the surveyed molding mills and timber kiosks (Source of data: Author's field survey)

	Worker background	Average number of workers
Molding mills (10 sites)	Immigrants[a]	2
	Owner's family	2
	Other	8
Total		12
Timber kiosks (15 sites)	Immigrants[a]	1
	Owner's family	2
	Other	1
Total		4

[a]Javanese, Minangkabau, Batak, and Melayu migrants came from Java, North Sumatra, and West Sumatra

Table 7.4 Monthly income and expenses of a Type 1 molding mill in October 2011 (Source of data: Author's field survey)

		Actual expense (Rp1000)	Percentage (%)
Sales		126,000	100
Cost of timber processing		99,514	79.0
Cost of material	Log initial price	76,600	60.8
Cost of labor (12 persons)	Log carrying	3150	2.5
	Molding mill operating	1680	1.3
	Timber waste cleaning	600	0.5
	Controlling, track driving, log loading	11,550	9.2
Other costs	Gasoline for the machines and tracks	4134	3.3
	Machine repairing	1400	1.1
	Land rental	400	0.3
Gross profit		26,486	21.0
Selling cost	Gasoline for tracks	106	0.1
Operating income		26,380	20.9
Operating income per day		854	
Capital	Band saws	150,000	

The mill's operating income was IDR 26,380,000 (approximately USD 1978). This value comprised approximately 21% of total sales. The owner said that the sales during this month were comparatively high because he had produced and sold timber to a public project. The mill sold a total of approximately 200 m³ of timber during the month.

7.3.2 Timber Supplies in Molding Mills and Timber Kiosks

7.3.2.1 Change in Species and Sizes of Logs and Timber

The research made clear that the species and sizes available in both molding mills and timber kiosks had changed in recent years. Table 7.5 details the 22 species available in the surveyed businesses. Based on interviews with managers, until a few years prior to the survey, Punah (*Teramerista glabra Miq.*), Kulim (*Sorodocarpus borneesis*), Meranti (*Shorea* spp.), and Tembesu (*Fagraea* spp.) were mainly sold for house construction, because these species are highly durable and not susceptible to insect damage. However, in recent years, the amount of logs of these species had decreased drastically. As a result, many molding mills and kiosks had difficulty buying these types of logs and timber. At the time of the survey, these species were

Table 7.5 Timber species processed and retailed in the surveyed molding mills and kiosks and their utilization (Source of data: Author's field survey)

No.	Local name	Scientific name	Utilization								
			Pulp pallet	Construction material							Wooden fitting and furniture
				Foundation (underground)	Foundation	Structure	Floor	Wall	Beam and Rafter		
1	Acacia	*Acasia cracicarpa/ mangium*	○	—	—	○	○	○	○	—
2	Balam	*Palaquium smatranum*	—	—	—	—	—	—	—	○
3	Bintangur	*Calophyllum* spp.	—	—	—	○	—	—	—	—
4	Daru–Daru	*Unknown*	—	—	—	—	—	—	○	—
5	Durian	*Durio carinatus*	—	—	—	○	—	○	○	—
6	Gedak	*Unknown*	—	—	—	—	○	○	—	○
7	Grongong	*Cratoxylon arborescens*	—	—	—	—	○	—	○	—
8	Jengkol	*Archidendron pauciflorum*	—	—	—	○	—	○	—	—
9	Karet	*Hevea brasiliebsis Muell. Arg*	—	—	—	○	—	—	○	—
10	Kempas	*Koompassia malacensis*	—	—	—	—	—	—	—	○
11	Kulim	*Sorodocarpus borneesis*	—	—	—	—	—	—	—	○
12	Mahang	*Macaranga* spp.	○	○	—	—	—	—	—	—
13	Meranti[a]	*Shorea* spp.	—	—	—	—	—	○	○	○
14	Marpoyang	*Unknown*	—	—	○	○	—	—	—	—
15	Pinang Janda	*Unknown*	—	—	○	○	—	—	—	—

(continued)

Table 7.5 (continued)

| No. | Local name | Scientific name | Utilization | | Construction material | | | | | | |
			Pulp pallet	Foundation (underground)	Foundation	Structure	Floor	Wall	Beam and Rafter	Wooden fitting and furniture
16	Pisang–pisang	*Mezzettia paviflora* Becc.	—	—	—	○	—	—	—	—
17	Punah	*Teramerista glabra* Miq.	—	—	—	—	—	—	—	○
18	Rengas	*Gluta* spp.	—	—	—	○	—	—	—	—
19	Sengon	*Paraserianthes falcataria*	○	—	—	—	—	—	—	—
20	Tembesu	*Fagraea* spp.	—	—	—	—	—	—	—	○
21	Telas	b	—	—	—	○	—	—	—	—
22	Ubar	*Eugenia* spp/*Syzigium* spp	—	—	○	—	—	—	—	—

[a]Meranti includes Meranti Merah, Meranti Kuning, and Meranti Batu
[b]Telas refers to fallen trees in the forest. The species retailed as Telas were Ubar (*Eugenia* spp. and *Syzigium* spp.), Kulim (*Sorodocarpus borneensis*), and Daru–Daru (unknown)

already recognized as rare and were used only for wooden fittings. Their prices were also continually on the rise.

Conversely, Acacia (*Acacia cracicarpa*), Mahang (*Macaranga* spp.), Karet (*Hevea brasiliebsis Muell. Arg*), and Sengon (*Paraserianthes falcataria*) had replaced Punah, Tembesu, and Kulim as the main timber species for house construction. These species grow quickly and produce timber in a short time span. They are not planted to be cut, but rather grow naturally in forests. Generally, during hardwood timber shortages, demand for these species was expected to spike. Acacia, Mahang, and Sengon were also essential materials for wooden pallets. The molding mills working as subcontractors of the pulp and paper companies specialized in processing these species. The buying price of Acacia, Sengon, and Mahang from the timber distributors was IDR 300,000 to 350,000 per m^3 (approximately USD 23–26). The buying price of timber for house construction ranged from IDR 1,000,000 to IDR 1,500,000 per m^3 (approximately USD 75–113).

Some timber kiosks also adapted their operations due to the hardwood shortage, for example, by selling a wider variety of species and buying any timber large enough in size for house construction from loggers and middlemen. In contrast, some timber kiosks concentrated on selling only a few timber species. For example, several timber kiosks sold only Mahang for house foundations, or sold thin Mahang logs, which were easy to obtain and requires no processing (local people place Mahang piles on the muddy ground to prevent houses from sinking). Such specialization indicates an adaptation to local timber supply and demand.

Although the industry classifies 22 timber sizes, the length of a single piece is typically 400 cm. By the time of the survey, however, some kiosks had started to sell hardwood timber in lengths of 200 cm. This could indicate a further adaptation; by shortening the length, kiosks were able to continue selling hardwood. Some timber kiosks had also begun to sell lumber. Lumber was cut roughly by hand saw rather than processed by band or circle saws in sawmills and molding mills. Although the lumber's surface was not smooth, the longer length could be maintained, with kiosks offering sizes of 5 W × 3D × 400 L (cm) and 6 W × 4D × 400 L (cm). Some kiosk managers indicated that local people preferred larger-sized lumber pieces to smaller-sized timber pieces.

7.3.2.2 Timber Supply

Based on interviews with managers, until several years ago, the molding mills and kiosks cooperated to share information such as timber distribution channels and prices, but no longer did so at the time of the survey due to the declining number of timber processing and retailing operations in Pekanbaru.

Distributors transported timber to the molding mills and timber kiosks either via the roads or the Kampar River. Table 7.6 depicts the sources of timber for the surveyed molding mills and timber kiosks according to producing district and forest type. The molding mills bought lumber produced in Siak, Kampar, and Pelawan. In Siak, the main areas of timber production were concession forests, which were

Table 7.6 Sources of timber in the surveyed molding mills and timber kiosks according to producing districts and forest type (Source of data: Aurthor's field survey)

	Districts of timber producing	Numbers by forest types of timber producing			
		Concession	Village forest	Unknown	Total
Numbers of molding mills by timber producing districts	Siak	3	1	0	4
	Kampar	0	4	0	4
	Pelalawan	0	2	0	2
	Total	3	7	0	10
Numbers of timber kiosks by timber producing districts	Siak	0	6	1	7
	Kampar	0	5	1	6
	Unknown	0	0	2	2
	Total	0	11	4	15

managed by the pulp and paper companies or their subcontracted mills. In Kampar and Pelawan, village forests were the main areas of timber production.

The timber kiosks bought timber that was processed in Siak and Kampar. Although there were several molding mills in Pekanbaru, the timber kiosks did not buy timber from them, because many molding mills in Pekanbaru processed mainly softwoods for wooden pallets and did not meet their demands. Some kiosk owners said that village forests were the main areas of timber production in both Siak and Kampar. Others did not know where timber was cut.

The author found that at the time of the survey, 22 out of 25 of the molding mills and timber kiosks were facing timber shortages. First, the hardwood supply was declining, as described in the previous section. Second, no one could bring timber to Pekanbaru without the necessary permits.

The survey revealed that there were three main types of timber supply chains in Pekanbaru, as depicted in Table 7.7. There were three stages of supply, namely, logging in forests, distributing logs to mills, and processing logs into lumber or timber in molding mills and, sometimes, in timber kiosks. Although kiosks had no machines to process the timber, kiosks managed timber production, with workers processing logs into lumber or timber in places other than the kiosks. In the first category, workers at the molding mills and timber kiosks managed all stages of the timber supply. This type of operation often managed large concessions and, for this reason, did not require many networks to secure adequate timber supply. The molding mills and timber kiosks that processed and retailed timber for wooden pallets fell into this category.

The second category involved timber distributors working together with molding mills and timber kiosks. The timber distributors often contacted loggers to keep abreast of availability and logging conditions. After preparing logs or lumber for transport, they called the molding mills and timber kiosks. The molding mills and timber kiosks paid half of the money needed for the distributors to buy the logs and covered the costs of transportation and labor. Upon receipt of the logs, they paid the remaining balance. In the case of timber kiosks, the bulk of the timber processing

Table 7.7 Timber supply chain types of the surveyed molding mills and timber kiosks (Source of data: Aurthor's field survey)

No.	Types of timber supply	Numbers of molding mills (Average numbers of networks)	Numbers of timber kiosks (Average numbers of networks)
1	Managing by themselves from logging to processing	5 (1)	3 (1)
2	Working with timber distributors	1 (2)	5 (3)
3	Buying timber from timber distributors	4 (3)	7 (3)

was conducted in a sawmill or molding mill near the forest. The surveyed molding mills worked with an average of two groups of distributors and the timber kiosks worked with three groups.

In the third category, loggers, distributors, and retailers worked together. The molding mills and timber kiosks bought timber that was delivered by the distributors. Some distributors were business partners of the mills or kiosks, while others visited the molding mills and timber kiosks without any prior contact to sell timber. Some managers noted that loggers sometimes brought timber directly to the molding mills and timber kiosks, bypassing distributors. The surveyed molding mills and timber kiosks had an average of three networks with timber distributors.

7.4   Discussion

This chapter investigated the socioeconomic features of the processing and retail sectors of the timber industry in Pekanbaru, focusing on the nature and scale of timber production, sales, and supply chains. Better understanding these features can inform the improvement of reforestation efforts by local communities.

A random survey and interviews of molding mills and timber kiosks in Pekanbaru found that some operations essentially function as subcontractors of bigger pulp and paper companies. This was because they could legally process and retail timber under the companies' existing permits without having to obtain any additional permit. These local businesses necessarily grow in line with the pulp and paper companies' business. The production process of pulp and paper companies has degraded the tropical forest; in response to monitoring and criticism of many institutes and non-governmental organizations (NGOs), the companies have begun producing timber from plantations and preserving the remaining forest in and around their concession area (Suzuki 2016). Previous research has concluded that the growth of pulp and paper companies depends on the level of forest conservation they practice in their business (Suzuki 2016). The molding mills and timber kiosks working for these companies will also need to evaluate the sustainability of their businesses and develop viable business strategies to ensure a stable local timber supply.

While some operations acted as subcontractors to bigger companies, other molding mills and timber kiosks sold timber directly to local people and government entities. They modified their businesses to adapt to emerging conditions, for example responding to timber shortages by changing timber species from hardwoods to softwoods, selling any kind of lumber to retain timber size, and so on. To alleviate the timber shortage, regulations regarding timber production and sales need to be reduced, especially regulations around the production and retail of softwoods that require prior approval.

The survey found that village forests represented an essential timber source for the molding mills and timber kiosks, which supplied timber to local government entities and local people. Village forests are granted forest management rights by the state and are often used in the context of social forestry, which is a sustainable forest management system managed by the state to improve the welfare, environmental balance, and socio-cultural dynamics at the local village level (Peraturan Pemerintah No. 23 Tahun 2021). More discussion is needed on how to connect such village forest management and local timber supply chains.

Successful reforestation by local communities requires developing stable income sources to support and sustain local forest management in the long term. Timber production and retail can be a way for local people to obtain income. To develop the timber sector in the context of local reforestation, it is therefore necessary to consider how local people connect with the local timber processing and retail sectors and how to improve local timber supply chains. In such a consideration, the social networks between molding mills and timber kiosks, which this study analyzed, should be recognized as essential. In this regard, reviving the local timber association may prove useful as a means to recognize and develop closer ties among the various sectors of the timber industry.

In recent years, local reforestation efforts have expanded in many areas in Riau, particularly in the context of rehabilitating degraded peatlands. Forests need to produce timber that fulfills local construction demands, including to supply appropriate sizes of building materials. Reforestation programs therefore must construct sustainable management systems that restore forests while at the same time using forest resources. This includes developing land ownership and use systems to produce timber sustainably and replanting tree species that are useful as timber.

Acknowledgments   My fieldwork was conducted with research permission (0343/SIP/FRP/SM/X/ 2011) from the State Ministry of Research and Technology (RISTEK), Indonesia. I am deeply grateful to my counterpart Dr. Bambang Subiyanto (LIPI). I am also grateful to Associate Professor Haris Gunawan (Riau University) and Ms. Erli Zani (at Andalas University at the time of my field research) for supporting my fieldwork. I also appreciate the cooperation of all the informants in the local molding mills and timber kiosks in Pekanbaru.

This research was supported by the Global COE Program "In Search of Sustainable Humanosphere in Asia and Africa" and the "Institutional Program for Young Researcher Overseas Visits," both of the Center for Southeast Asian Studies, Kyoto University.

References

Barnard TP (1998) The timber trade in pre-modern Siak. Indonesia 65:86–96. https://doi.org/10. 2307/3351405

BPS (Badan Pusat Statistik) (2012) Statistik Indonesia 2012. Badan Pusat Statistik, Jakarta

BPS Provinsi Riau (2019) Pertumbuhan Ekonomi Riau Tahun 2019. https://riau.bps.go.id/ pressrelease/2020/02/05/700/pertumbuhan-ekonomi-riau-tahun-2019.html.   Accessed 10 Jan 2022

BPS Provinsi Riau (2022a) Penduduk Kabupaten/Kota (Jiwa) 2010–2011. https://riau.bps.go.id/ indicator/12/32/5/penduduk-kabupaten-kota.html. Accessed 10 Jan 2022

BPS Provinsi Riau (2022b) Penduduk Kabupaten/Kota (Jiwa) 2015–2017. https://riau.bps.go.id/ indicator/12/32/3/penduduk-kabupaten-kota.html. Accessed 10 Jan 2022

BPS Provinsi Riau (2022c) Penduduk Kabupaten/Kota (Jiwa) 2018–2020. https://riau.bps.go.id/ indicator/12/32/2/penduduk-kabupaten-kota.html. Accessed 10 Jan 2022

BPS Provinsi Riau (2022d) Penduduk Kabupaten/Kota (Jiwa) 2021–2023. https://riau.bps.go.id/ indicator/12/32/1/penduduk-kabupaten-kota.html. Accessed 10 Jan 2022

Casson A, Obidzinski K (2002) From new order to regional autonomy: shifting dynamics of "illegal" logging in Kalimantan, Indonesia. World Dev 30(12):2133–2151. https://doi.org/10. 1016/S0305-750X(02)00125-0

Iwanaga S, Masuda M (2012) Jawa ni okeru mokuzai kako kigyo wo chushin toshita jumin ringyo keiei kakuritsu heno kokoromi (Trail for local people's forestry establishment in partnership with timber processing company in Indonesia). Jpn J Int For For 83:42–47. https://doi.org/10. 32205/jjjiff.83.0_42

Iwanaga S, Hiratsuka M, Yaginuma H (2014) Indoneshia ni okeru Melaluca (*Meraleuca cajupti*) zai shijo no genjo to shinkisannyu kanosei: REDD+ purojekuto ni okeru chiikijumin no atarashi shunyugen no soshutsu ni mukete (Circumstances and possibility of entry into Melaleuca (*Meraleuca cajupti*) timber market in Indonesia: toward generation of new income source for local people in the REDD+ project). J For Econ 60(3):25–36. https://doi.org/10.20818/jfe.60.3_ 25

Kementerian Lingkungan Hidup dan Kehutanan (2019) Statistik lingkungan hidup dan kehutanan tahun 2018. Pusat Data dan Informasi Kementeritan Lingkungan Hisup dan Kehutanan, Jakarta

Kementerian Lingkungan Hidup dan Kehutanan (2020) Rencana Strategis Tahun 2020–2024. https://www.menlhk.go.id/site/single_post/3298. Accessed 10 Jan 2022

McCarthy JF (2006) The fourth circle: a political ecology of Sumatra's rainforest frontier. Stanford University Press, Stanford

Obidzinski K (2005) Illegal logging in Indonesia: myth and reality. In: Resosudarmo BP (ed) The politics and economics of Indonesia's natural resources. ISEAS Publications, Singapore, pp 193–205. https://doi.org/10.1355/9789812305497-018

Pastor G (1927) De Panglongs. Landsdrukkerij, Welthervreden

Peraturan Pemerintah No. 23 Tahun (2021). https://peraturan.bpk.go.id/Home/Details/161853/pp-no-23-tahun-2021. Accessed 10 Jan 2022

Potter L, Badcock S (2004) Tree crop smallholders, capitalism, and *adat*: studies in Riau Province, Indonesia. Asia Pacific Viewpoint 45(3):341–356. https://doi.org/10.1111/j.1467-8373.2004.00245.x

Suzuki H (2016) Indoneshia ni okeru kami purupu kigyo ni yoru shinrin hozen no torikumi: jisshi katei ni okeru kigyo to NGO no kankei (Forest conservation efforts by pulp and paper company in Indonesia: the relationship between a company and NGO). J For Econ 62(1):52–62. https://doi.org/10.20818/jfe.62.1_52

van Noordwijk M, Suyanto S, Budidaesono S et al (2007) Is Hutan Tanaman Rakyat a new paradigm in community based tree planting in Indonesia? ICRAF Working Paper no 45. ICRAF Southeast Asia, Bogor

Chapter 8
Toward Climate Change Mitigation: Restoration of the Indonesian Peat Swamp

Haris Gunawan, Dede Hendry Tryanto, Kosuke Mizuno, and Osamu Kozan

Abstract Indonesia created a breakthrough in peatland management by establishing the Peatland Restoration Agency in early 2016 with the aim of restoring 2.67 million ha of degraded peatlands. This effort is intended to accelerate the recovery of peatlands and return of its hydrological functions after extensive damage by fire, drainage canals, and other external factors. This paper highlights the potential biomass and carbon resources in various land-use covers located in the Riau Biosphere Reserve. It discusses the results of restoration experiments conducted in severely degraded peatlands, and estimates carbon emission reductions in targeted priority areas. The total estimated emission reduction in natural forests was higher than in logged-over forests and disturbed forests: 207.36 CO_2 Mg h^{-1}, 161.48 CO_2 Mg h^{-1}, and 65.87 CO_2 Mg h^{-1}, respectively. The restoration of 2.3 million ha of targeted peatland ecosystems was estimated to have reduced carbon emissions by 98.77–153.53 Mt CO_2e. The value of carbon from peatlands is considered important for maintaining ecological function while optimizing economic benefits. We have confirmed that above ground carbon storage can be restored even in severely degraded peatlands. Avoiding vegetation loss is an important aspect of restoration activity, but recovery of vegetation in degraded areas depends on below-ground carbon stocks, as these are indicative of fertile soils in various kinds of land cover and use.

Keywords Climate change mitigation · Emission reduction · Peat swamp ecosystem restoration

H. Gunawan (✉)
University of Riau, Pekanbaru, Riau, Indonesia

D. H. Tryanto
The Indonesian Environment Fund, Jakarta, Indonesia

K. Mizuno
School of Environmental Science, University of Indonesia, Jakarta, Indonesia

Center for Southeast Asian Studies, Kyoto University, Kyoto, Japan

O. Kozan
Center for Southeast Asian Studies, Kyoto University, Kyoto, Japan

© The Author(s) 2023
K. Mizuno et al. (eds.), *Vulnerability and Transformation of Indonesian Peatlands*, Global Environmental Studies, https://doi.org/10.1007/978-981-99-0906-3_8

8.1 Introduction

Communities around the world are increasingly aware of the fact that the steady rise of anthropogenic greenhouse gas (GHG) emissions in the atmosphere is a major cause of contemporary climate change. Mitigating GHG emissions to prevent a global rise in temperature above 1.5–2 °C is a major challenge, especially GHG emissions until 2030 are projected to be 52–58 Gt CO_2e. Increasing global temperatures have already begun to damage human life through extreme heat events, increasing sea levels, increases in drought, reduction of crop yields, and the destruction of the coral reef.

Agriculture and forestry, including changes in land use, are responsible for a quarter of the global GHG emissions (Masson-Delmotte et al. 2019). In Indonesia, almost half of total national emissions are related to forestry and peatland issues (MoEF 2018). Peatlands are huge deposits of organic carbon, and the alteration of their natural condition can lead to the release of stored carbon in amounts that may affect the global climate. Addressing global climate change requires transformative actions from multiple stakeholders and targeted programs that can be implemented consistently, transparently, and collaboratively.

Tropical peat swamp forests play an important role in the global carbon cycle, have tangible and intangible ecological value, and provide numerous environmental services (Yu et al. 2010; Page et al. 2011; Novita et al. 2020). A wise and sustainable management system is essential to optimize the economic, social, and ecological value of tropical peatlands (Rieley et al. 2005). Good peatlands management involves harvesting renewable resources sustainably while conserving nonrenewable resources and maintaining the attributes and functions of the peatlands.

Nearly half of the tropical peatlands in Southeast Asia, as much as 14.9 million ha, are found in Indonesia. Peat management requires improved knowledge of the carbon cycle in tropical peatlands and how it responds to land-use and land-cover change. Optimum carbon storage in tropical peatlands requires a combination of high vegetation biomass (carbon sequestration potential), a water table that is near to or above the peat surface for most of the year, and a slow rate of organic matter decomposition. Drainage of peatlands and other disturbances lead to increased aeration, decomposition, and carbon losses in surface peat (Hooijer et al. 2012; Page et al. 2011). Furthermore, peat fires related to land clearing occur frequently, releasing significant amounts of carbon from the forest vegetation and soil. Conversion of peatlands for agriculture and plantations requires radical changes in the vegetation cover and permanent drainage. These changes lower the peat water table, reducing or, in most cases, eliminating the natural capacity of the peatland system to sink carbon. Aeration leads to continuous aerobic decomposition of organic matter and increased peat temperatures, which then increases aerobic heterotrophic respiration and peat surface CO_2 emissions (Hooijer et al. 2009; Hooijer et al. 2008).

Even though the importance of peat swamp forest for carbon sequestration and other ecosystem functions is widely recognized they are one of the most rapidly disappearing forest types in the world (Hooijer et al. 2008). In a natural state, peat swamp forests are characterized by dense forest vegetation, thick peat deposits (up to 20 m), and a groundwater table that is above or close to the peat surface throughout the year (Hirano et al. 2009; Page et al. 2004; Takahashi et al. 2002). Peat has low bulk density (approximately 0.1 g cm^{-3}) as it is composed of approximately 10% tree remains and 90% water (Hooijer et al. 2010) and is 50–60% carbon by dry weight (Page et al. 2011; Neuzil 1997). Peat swamp forests also store large amounts of carbon in living plant biomass, with typical values ranging from 100 t C ha^{-1} to 250 t C ha^{-1} (Novita et al. 2020; Page et al. 2011).

Peatlands with 2772 t C ha^{-1} are based on an average peat thickness of 5.5 m (Page et al. 2011). The carbon reserve of these peatlands is very high, ranging from 30 kg C m^{-3} to 70 kg C m^{-3} (Agus et al. 2009), equivalent to 300–700 Mg C m^{-1} of soil depth. Initial estimates show that on average, peat in Sumatra is thicker and has a carbon content of approximately 3000 Mg ha^{-1} (Wahyunto et al. 2010). Data on biomass and carbon storage in the remaining peat swamp forests are scarce, however (van der Meer and Verwer 2011).

One of the main problems related to sustainable management of peat swamp forests is their current state of severe degradation. In the Riau Biosphere Reserve, land conversion and poor management has caused the loss of around 300,000 ha of natural peat swamp forest in the last 17 years. The uniqueness of this biosphere reserve is that it is a vast landscape consisting of a hydrological network of small lakes and streams and peat swamp forests. The dominant natural ecosystems are peat swamp forests, which are surrounded by land that is being used in various ways, e.g., production forests, degraded/abandoned lands, industrial plantations (timber and oil palm), and agricultural lands and settlements. The remaining natural peat swamp forest, located in the core area, consists of 84,967 ha in the Giam Siak Kecil Wildlife Reserve and 21,500 ha in the Bukit Batu Wildlife Reserve. The function of the core area is to conserve biodiversity, while the buffer zone functions to protect this area. The outer and largest area of the biosphere reserve functions as a transition area.

Levels of land degradation can be classified as severe, moderate, and heavy, with "severely degraded" characterized as the absence of trees or dead trees and open areas colonized by fern or grass. A "heavily degraded" forest has lost much of its original biodiversity and most of its structure (bintangur stands). These areas are colonized by grass, fern, and *Melastoma* sp. after an earlier succession. A "moderately degraded" forest regenerates naturally, and some residual trees still remain (e.g., *Palaquium sumatranum*). Natural regeneration of tree species in the peat swamp forests in the Riau Biosphere Reserve is vigorous, but some of the typical upper-story species (i.e., *Shorea* spp., *Gonystylus bancanus, Tetramerista glabra, Durio carinatus, Dyera lowii,* and *Calophyllum lowii*) show limited or even no regeneration. Rehabilitation is therefore an urgent ecological matter, particularly for maintaining and preserving species populations (Gunawan et al. 2012). Additionally, in the case of biosphere reserves, conservation and sustainable restoration efforts should be promoted in order to restore large areas of degraded peatlands and forests due to high fire intensity, poor canal drainage, and illegal logging activities; to reduce forest encroachment and conversion; to protect the livelihoods of local people; and to restore carbon sequestration and protect the remaining peat carbon stocks.

Restoration activities have attracted extensive attention in Southeast Asia in the last decade due to their potential to restore the multiple functions of the peat swamp ecosystem. There is limited information regarding the extent of these activities, however. Peatland restoration is still in its early stages and has been limited to experimental rehabilitation projects that aim to restore the main functions of peat ecosystems, especially the hydrological and carbon cycles (Page et al. 2008; Giesen 2004). The Government of Indonesia established the Peatland Restoration Agency (*Badan Restorasi Gambut*, BRG) in 2016 to reduce conversion of peatlands to other uses and accelerate the recovery and return of hydrological functions of peat that have been mainly damaged by fire and drying.

Protecting and managing peatland ecosystems can be done by avoiding the conversion of peatland to other land uses and enhancing peat soils capacity to storck carbon. This paper highlights potential biomass and carbon resources in various land-use covers located in the Riau Biosphere Reserve and current results of restoration experiments in severely degraded peat swamp ecosystems.

8.2 Methods

8.2.1 Data Collection

8.2.1.1 Above-Ground Carbon Storage

Our study observed above-ground biomass in 0.5 ha from a 25×25 m^2 plot for natural forest (plot 1, plot 2, plot 3), logged-over forest (plot 4), and wind-disturbed forest (plot 5, plot 6). For each plot, the authors determined the structure and species composition of trees that had a diameter at breast height (dbh; 1.3 m) of over 3 cm. Allometric equations were applied to estimate biomass from dbh as a dependent variable.

Natural peat swamp forests, which are mostly found in biosphere reserves where there have been no tree-felling activities and no drainage canals, can be classified into two main forest types: mixed-peat swamp forest and bintangur (*Callophylum lowii*) forest. These forest types are distinguished by dominant tree species, vegetation communities, floristic composition, and basal areas. We found high species richness, tree diversity, and unique environmental characteristics in both types of peat swamp forest. Mixed-peat swamp forest located in the upstream river basin of the conservation area was identified by different environmental characteristics, such as the water situation, floristic composition, basal areas, and density. The bintangur forest was located far away from the river basin; *Callophylum lowii* (local name:

bintangur) is the dominant species. Bintangur trees have unique vegetation formations and exist in very deep peat (>10 m).

In logged-over forest these dominant tree species of the peat swamp forests have been felled. Natural regeneration does not occur for almost 10 years after logging activities end. In 2005, a large peat swamp forest area was burned and then colonized by ferns and *Melastoma* spp. The landscape became more open, allowing the wind to attack the forest stands. Wind-disturbed forest areas are surrounded by an acacia timber plantation and have been designated a forest conservation area by the forest company. At the same time, however, the open landscape has become more accessible to people, and the population of the nearby village has increased dramatically since 2008.

Improved rehabilitation management is needed to reestablish the peat swamp forest ecosystem dominant tree species and vegetation communities. The main causes of degradation in logged-over forest were logging and forest conversion. Since the early 2000s, illegal logging activities have occurred in those areas, and approximately 3 km on both sides of the Bukit Batu river basin were degraded forest areas. The level of degradation varies between severely, heavily, and moderately degraded. The severely degraded areas were characterized as being mostly covered by grass and fern. Heavily degraded areas were covered by *Macaranga* sp. or woody pioneer species, and moderately degraded areas had regenerating woody species that formed the main vegetation of peat swamp forests.

Our study found two main differences in the logged-over forest and wind-disturbed forest. First, in relation to the regeneration performance of the upper-story tree species, we found that *Palaquium sumatranum* and *Callophlum lowii* have vigorous recovery capacity. Our study therefore indicates that recovery is possible in both logged and wind-disturbed peat swamp forests in the Riau Biosphere Reserve, where *Palaquium sumatranum* and *Callophylum lowii* are the dominant species. We found secondary forest recovery even when most of the upper story was absent due to limited or even no regeneration.

8.2.1.2 Below-Ground Carbon Storage

At each site, natural forest (NF), logged-over forest (LOF), wind-disturbed forest, oil palm plantations, and rubber garden soil were sampled through the use of a random sampling design. A peat auger was used to determine peat depth. Peat samples were taken from each corner and center of a square sampling area (25 m^2) from the surface to a depth of 40–60 cm to determine the percentage of carbon content and bulk density. Following the bulking of samples from each area, the samples were transported to the Soil Laboratory of Bogor Agriculture University, where they underwent analysis upon arrival.

8.2.2 *Analysis*

8.2.2.1 Biomass and Carbon Storage

Total above-ground biomass in each plot was estimated using an allometric equation (Hiratsuka et al. 2006; Brown 1997):

$$Y = \exp.(-2.134 + 2.53^* \ln (D)) \tag{8.1}$$

where Y is total above-ground biomass in kg tree^{-1} and D is diameter at breast height (dbh in cm).

For comparison, we used the allometric equation from the following studies below:

$$Y = \exp.(-2.289 + 2.694 \ln (D) - 0.021 (\ln (D)) \tag{8.2}$$

$$BP \text{ (above– ground biomass)} = 0.1236 \, D^{2.3677} \tag{8.3}$$

$$W \text{ (stem of biomass)} = 0.153108 \, 0 \, (D)^{2.40} \tag{8.4}$$

$$W \text{ (stem of biomass)} = 42.69 – 12.800 \, (D) + 1.242 \, (D^2) \tag{8.5}$$

$$W \text{ (stem of biomass)} = 0.2902 \, D^{2313} \tag{8.6}$$

Equations (8.4, 8.5, and 8.6) were used to calculate the biomass of trees planted in experimental plots in forested or burned areas.

Above-ground carbon storage was calculated by assuming that carbon storage was 0.5 of the total above-ground biomass (Brown and Lugo 1982). The team used other secondary data sources to estimate the amount of above-ground carbon storage in acacia, rubber, and oil palm plantations. These data were used to determine the total amount of above- and below-ground carbon storage for all different land uses in the Riau Biosphere Reserve.

8.2.2.2 Below-Ground Carbon Storage

The estimation of below-ground carbon storage (Mg C ha^{-1}) involved determining the bulk density of the soil's organic matter content. Bulk density was determined by the tube core or clod method with the following formula (Murdiyarso et al. 2005):

$$\begin{aligned} &\text{Bulk density } \left(\text{g cm}^{-3}\right) \\ &= (\text{weight of soil with tin} - \text{weight of empty tin})/\text{volume of the tin} \end{aligned} \tag{8.7}$$

The below-carbon storage was estimated using the following equation:

$$C \text{ (ton)} = \sum [A \text{ (each depth category)} \times D \text{ (in each category)}] \times BD \times CC \quad (8.8)$$

where
   C: total carbon (Mg C)
   A: area (ha)
   D: average depth (meter)
   BD: bulk density (g cm^{-3})
   CC: carbon content in peat sample (%)

8.3 Results and Discussion

Above-ground biomass was calculated for the different forest types to indicate the relative proportions of carbon storage and carbon sequestration (see Fig. 8.1). Mean carbon storage per hectare varied from 10.6 Mg C ha^{-1} to 60.8 Mg C ha^{-1}. Comparisons using other equations showed that carbon storage was high in all of the sampling plots (Figs. 8.1 and 8.2). The highest rates of carbon storage was found in the natural forest (plots 1, 2, 3), followed by logged-over forest (plot 4) and wind-disturbed forest (plots 5 and 6). The highest amount of stored carbon ranged from 89.6 Mg C ha^{-1} to 98.2 Mg C ha^{-1} in natural forest (plots 1, 2, 3). The estimated above-ground carbon stock in the natural forest plots was slightly lower than in the forest sampled in the same area by Istomo (2002), who reported mean above-ground C stocks of 131 Mg C ha^{-1} in Riau Province. We found logged-over forest stocks

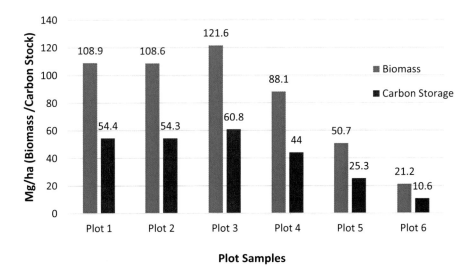

Fig. 8.1 Above-ground biomass and carbon storage

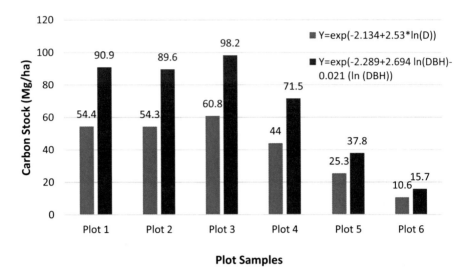

Fig. 8.2 The amount of carbon storage in above-ground biomass estimated by two different allometric equations

71.5 Mg C ha^{-1}, while wind-disturbed forest carbon stock (plots 5 and 6) ranged from 15.7 Mg C ha^{-1} to 37.8 Mg C ha^{-1}.

The above-ground biomass ranged from 108.9 Mg ha^{-1} to 121.6 Mg ha^{-1} for natural forest and dropped to 88.1 Mg ha^{-1} for logged-over forest and 21.2 Mg ha^{-1} for wind-disturbed forest. A sharp drop in biomass occurred in disturbed forests, especially those affected by wind and fires. This shows that natural disturbances negatively impact the protection of the biomass and carbon storage of the remaining peat swamp forests in the Riau Biosphere Reserve. Land-cover changes from natural vegetation to industrial forest plantations also have a negative effect. Wind-disturbed forest is one of the remaining natural forest types in the buffer zone of the Riau Biosphere Reserve.

Based on the below-ground carbon stock analysis of peatlands in the Riau Biosphere Reserve, logged-over forest has the highest value (5981 Mg C ha^{-1}), followed by acacia plantations (5460 Mg C ha^{-1}), and natural forest (4200 Mg C ha^{-1}; Fig. 8.3). The lowest below-ground carbon stocks were found in oil palm plantations (3960 Mg C ha^{-1}).

The results show higher below-ground peatland carbon storage in both forested and developed peatland areas compared to the results of Agus et al. (2009) and Wahyunto et al. (2010). High variation in below-ground carbon is also reported, with amounts ranging from 39 kg C m^{-3} to 66 kg C m^{-3}, which is equivalent to 390–660 Mg C ha^{-1} m^{-1} of soil depth. Initial estimates show that on average, peat in Sumatra is thicker and has a carbon content of approximately 3000 Mg C ha^{-1}, while the peat in Kalimantan is thinner and contains about 2000 Mg C ha^{-1}. In comparison, the carbon content of mineral soil is usually concentrated in the first

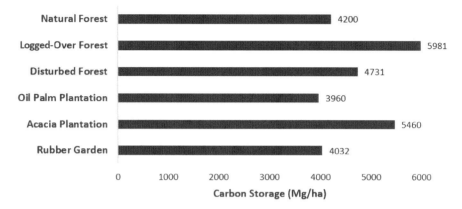

Fig. 8.3 Below-ground carbon storage for various land uses

several centimeters of the surface layer and rarely exceeds 250 Mg C ha^{-1} (Wahyunto et al. 2010).

Above-ground biomass in tropical forests may vary considerably in response to differences in climate and soil parameters. For example, a study in Rondonia, Western Brazil, found that above-ground biomass in primary forest was 290–495 Mg ha^{-1} and about 40–60% less in young secondary forests in the same area. Another study in Sumatra showed that dipterocarp hill forest had an above-ground biomass ranging from 271 Mg ha^{-1} to 478 Mg ha^{-1}, whereas the above-ground biomass of other lowland dipterocarp forest was 509 Mg ha^{-1}. According to values published in various sources, the above-ground biomass carbon stored in undisturbed peat swamp forest ranges from 73 Mg C ha^{-1} to 323 Mg C ha^{-1}, while that of logged-over and secondary peat swamp forest was between 65 Mg C ha^{-1} and 167 Mg C ha^{-1} (van der Meer and Verwer 2011).

In Central Kalimantan, the mean total ecosystem carbon stock of primary and secondary peat swamp forests was 1770 ± 123 Mg C ha^{-1} and 533 ± 49 Mg C ha^{-1}, respectively (Novita et al. 2020). Another study from West Kalimantan reported that the mean total ecosystem carbon stock was 4021 Mg C ha^{-1}, with peat thickness ranging from 650 cm to 1050 cm (Basuki 2017). The estimated amount of above-ground biomass was 252 Mg ha^{-1} in the Riau natural peat swamp forest, 111.4 Mg ha^{-1} in the logged-over forest, and 65 Mg ha^{-1} in the secondary forest. In comparison with other types of forest, logging forest in Jambi, Sumatra, resulted in a biomass decrease of 158.8 Mg ha^{-1}. The total above-ground biomass can be influenced by soil fertility, and the occurrence of different forest types depends on the hydrology, chemistry, and organic content of the peat. Tropical peat swamp forests are characterized by high acidity (pH < 4), low nutrient quality, and a low rate of litter input (Neuzil 1997).

The total amount of above-ground carbon storage in lowland peat swamp forest in Indonesia was around 230 Mg C ha^{-1}, with below-ground carbon storage of 2425 Mg C ha^{-1}. Peat swamp forest stores large amounts of carbon in plant biomass,

Table 8.1 Estimation of average below-ground carbon storage, location, and mean thickness (Novita et al. 2020; Page et al. 2011; Wahyunto et al. 2010)

Location	Thickness (m)	Carbon storage (Mg C ha^{-1})
Kalimantan	<3	2310
Sumatra	>5.5	3000
Riau Biosphere Reserve (study area)	6.6	4727
Brunei	7	3528
Malaysia	7	3528

with typical values ranging from 100 Mg C ha^{-1} to 250 Mg C ha^{-1} (Page et al. 1999). Below-ground carbon storage on peatlands with 2772 Mg C ha^{-1} is based on an average peat thickness of 5.5 m (Page et al. 2011). Peat in Sumatra is thicker and has carbon storage of approximately 3000 Mg ha^{-1} (Wahyunto et al. 2010).

The results of this study show that the amount of above-ground carbon storage varies according to forest type and land use. The intensity and different causes of forest degradation affect how the vegetation stores and sequesters carbon. Forest degradation caused by humans could have less impact compared to natural disturbances, such as frequent fires and wind, which slow down the natural recovery of forests. As Binkley et al. (1997) stated, disturbances in natural forests may cause regeneration delays because viable seed sources are not available and the site is not suitable for the recovery of forest vegetation. The above-ground carbon storage in the peat swamp forests that remain today may vary greatly due to disturbances and natural variation (Verwer and van der Meer 2010).

The decrease in the amount of above-ground carbon between natural forest and logged-over and wind-disturbed forest highlights a decrease or malfunction in carbon sequestration following land-cover and land-use changes as a result of fire or the conversion to agricultural land. The same pattern is observed when comparing the amount of biomass in the different types of peat swamp forest.

Most of the research area in the Riau Biosphere Reserve is covered by deep peat ranging from 4.5 m to 10 m in depth that has carbon stock of 3960–5981 Mg C ha^{-1}. This is higher than the amount of below-ground carbon stored in other peat areas in other Southeast Asian countries, including other areas in Indonesia such as Kalimantan (Novita et al. 2020) and even Sumatra (Page et al. 2011; Wahyunto et al. 2010; see Table 8.1).

Despite differences in biomass and carbon sequestration capacity, these forest areas play a crucial role as an essential pool of carbon. In terms of reducing carbon losses from land-cover changes and degradation, the rehabilitation of degraded peat swamp forests was recommended as the preferred land-use policy for future development in the Riau Biosphere Reserve. The results of this study suggest that degraded peat swamp forests currently contain between 15.7 Mg C ha^{-1} and 37.8 Mg C ha^{-1} and have the potential to increase their above-ground carbon content to at least 89.6–98.2 Mg C ha^{-1}. Consequently, conservation and restoration of the peat swamp forests is crucial for conserving and even improving the amount of carbon stored in the Riau Biosphere Reserve.

Table 8.2 Above- and below-ground carbon storage in different land uses

Land-use cover	Above-ground carbon storage (Mg C ha^{-1})	Below-ground carbon storage (Mg C ha^{-1})	Percentage of carbon	Bulk density (g/cc)	Peat depth (m)
Natural forest	94.25	4200	0.56	0.125	6
Logged-over forest	71.5	5981	0.55	0.145	7.5
Disturbed forest	26.75	4731	0.57	0.083	10
Rubber garden	68[a]	4032	0.56	0.16	4.5
Acacia plantation	42.57[a]	5460	0.56	0.15	6.5
Oil palm plantation	42.3[a]	3960	0.55	0.12	6.5

[a]Based on other studies: rubber (Palm et al. 2004); acacia (Masripatin et al. 2010); oil palm (Murdiyarso et al. 2010)

Some typical canopy species of peat swamp forest are preferred in the rehabilitation of logged-over forest areas. For example, *Tetrameristra glabra, Palaquium burckii,* and *Xylophia havilandii* grow well in more open-gap plots. In addition, keeping wet conditions on the forest floor will encourage continuous growth. These results show that level degradation of the peat swamp forest ecosystem and selection of suitable tree species would influence the success of vegetation rehabilitation efforts. Table 8.2 shows that the amount of carbon is generally much higher below ground for all types of land use.

Table 8.2 highlights the importance of the peat swamp forest ecosystem for all carbon storage, not just the carbon stored in trees. This should be taken into consideration when evaluating management options for the Riau Biosphere reserve. The largest amount of below-ground carbon of 5981 Mg C ha^{-1} was found in logged-forest areas, followed by peatlands under acacia plantations (5775 Mg C ha^{-1}). The lowest amount of below-ground carbon was 3960 Mg C ha^{-1} and was found in oil palm plantations.

Natural forest had the highest amount of above-ground carbon, followed by logged-over forest. The lowest was wind-disturbed forest. In developed peatlands, the highest amount of carbon was measured under rubber trees and was 68 Mg C ha^{-1}, while *Acacia crassicarpa* had 42.57 Mg C ha^{-1} and *Elaeis guineensis* had 42.3 Mg C ha^{-1}. These differences could be useful for addressing concerns about maintaining carbon storage in the peatland ecosystem when managing these plantations or choosing land uses.

The percentage of carbon composition is similar in forested areas and developed peatland areas. In contrast, bulk density was the highest in rubber gardens compared to other the sample plots and lowest in disturbed forest areas. Bulk density can be determined by the degree of decomposition, and the generally low bulk density values indicate low decomposition in peat. Rubber gardens had the highest bulk density because the peat had decomposed in advance compared to the other plot areas.

The bulk density of peat soil is likely the most important parameter because it varies considerably across peat types and even within peat types. Fibric tropical forest peats have a bulk density of less than 0.1 g cm^{-3}, while highly decomposed sapric peats have values in excess of 0.2 g cm^{-3}. Higher values are common in areas with topogenous clayey peats near the fringes of dome-shaped peat formations in areas with intensive agriculture management, such as paddy fields and oil palm plantations. The bulk density values of surface soils in the central dome areas under peat forests are generally lower than those peats under fringe domes of mixed swamp forest vegetation. In shallow fringe areas of peat domes where peat soils commonly contain clay, bulk density values are higher at certain depths than in the surface soils (Driessen and Rochimah 1976).

The bulk density of peatland under natural forest, logged-over forest, and disturbed forest means it can be classified as fibric and hemic, with bulk densities ranging from 0.083 g cm^{-3} to 0.145 g cm^{-3}. Meanwhile, the carbon density of these same forest types ranges from 4200 Mg C ha^{-1} to 5981 Mg C ha^{-1}. In developed peatland areas, the bulk density ranges from 0.12 g cm^{-3} to 0.16 g cm^{-3}, classifying them as hemic. The carbon density of peatland under development in rubber gardens, acacia plantations, and oil palm plantations ranges from 3960 Mg C ha^{-1} to 5460 Mg C ha^{-1}.

Generally, there are patterns in the carbon storage of peat in the Riau Biosphere Reserve. For example, forested areas have higher carbon compared to developed peatland areas, mainly due to peat decomposition processes. In contrast, developed peatland areas that had a continuously managed hydrology regime (e.g., forested areas and acacia plantations) showed similar amounts of below-ground carbon storage. Insufficient hydrology management can lead to lower amounts of below-ground carbon, however (e.g., oil palm plantations and rubber gardens). Therefore, management of peatlands significantly influences bulk density and the rate of the decomposition process.

8.3.1 Biomass Restoration and Carbon Reduction

Increase in biomass, carbon storage, and carbon sequestration in forested areas occurs through natural regeneration processes. Restoration activities with revegetation and natural forest regeneration approaches in tropical peatland ecosystems are effective and natural climate solutions for emissions reduction. The potential for emissions reduction through revegetation and natural regeneration shown in this study provides evidence that avoiding vegetation loss is an important part of restoration activities. The total estimated emission reduction provided by natural forest was higher than that estimated for logged-over forest and wind-disturbed

Fig. 8.4 Existing revegetation area in Tanjung Leban village, Riau Province, 10 years after the integrated restoration measures (photo by H. Gunawan)

forest: 207.36 CO_2 Mg h^{-1}, 161.48 CO_2 Mg h^{-1}, and 65.87 CO_2 Mg h^{-1}, respectively.

The Research and Development Division of the Peatland Restoration Agency conducted action research between 2016 and 2020 that was based on robust science and technology. A pilot project of revegetation with native tree species, or paludiculture, was mixed with the agroforestry model to maximize land utilization and improve the livelihoods of local people. The pilot project and paludiculture research were developed in degraded peatlands, and the results showed that revegetation can turn degraded peatlands into cultivation areas that improve the livelihoods of local people while mitigating peat fires and improving carbon sequestration.

Based on the authors' study in South Sumatra, revegetation in burned peat areas can have a positive impact on the sequestering of carbon in belangeran (*Shorea balangeran*), jelutung (*Dyera lowii*), and ramin (*Gonystylus bancanus*) stands. A combination of jelutung (*Dyera lowii*) and ramin (*Gonystylus bancanus*) has a potential carbon sequestration of about 35.82 t CO_2 ha^{-1} $year^{-1}$ at 6 years of age and 56.5 t CO_2 ha^{-1} $year^{-1}$ at 8 years of age in belangeran (*Shorea balangeran*) stands (Bastoni 2019). The success of above-ground biomass recovery by doing a participatory restoration action research is in 2.5 ha plot of degraded peatland areas during almost 14 years in Tanjung Leban Village, Riau Province (Fig. 8.4). This plot has been extended by an integrated restoration actions were rewetting, revegetation, and revitalization of livelihood in a total of 2400 ha of water zones. The above-ground carbon sequestration has been restored even in heavily degraded peatlands. The Tanjung Leban plot is a landmark in tropical peatland restoration and has the potential to become a model for the permanent restoration of the peatland landscape and promote peatland green economy in the near future.

8.3.2 Potential Carbon Credit in the Peatlands Ecosystem

Based on the Indonesia Peatland Reference Emissions Level, the historical annual emissions are estimated to be 10.8 Mt CO_2 year^{-1} and will reach up to 612 Mt CO_2 by 2030. The Peatland Restoration Agency (Stage 1) has applied rewetting to degraded peatlands while improving community livelihoods in an effort to reduce carbon emissions. This program is expected to reduce emissions to 305.6 MtCO_2 in 11.6 million ha located in prioritized restoration areas. The main restoration target will be 2.3 million ha that will contribute up to 54–74% or 98.77–153.53 MtCO_2e of the emissions reduction (CCROM IPB 2021). The value of carbon from peatlands is important for maintaining ecological function while optimizing economic benefits. Community awareness regarding the importance of conserving the remaining peat forests and restoring degraded peatlands should be improved as these lands act as frontiers for this rich carbon ecosystem. Payment for Ecosystem Services (PES) through paludiculture may be key to increasing carbon sequestration and reducing economic pressure.

8.4 Conclusion

In this study, revegetation with native species contributed to climate change mitigation strategies by enhancing carbon sequestration, so reducing the amount of CO_2 in the atmosphere. Avoiding peat deforestation and degradation during revegetation activities would increase potential emissions reduction provided by the peatland ecosystem. Human activities and natural disturbances reduce the ability of the peat swamp forest ecosystem to sequester CO_2, as shown by our analysis of logged-over and wind-disturbed forest. Improved management of secondary forest must be achieved through rehabilitation, ceasing forest conversion, and reducing the impact of wind and fire disturbances. Natural regeneration is very important to improving the condition of secondary degraded peat swamp forest, but is not sufficient in itself for recovery of forest vegetation and associated biodiversity. At this point, some form of human-assisted regeneration is needed.

The indigenous tree species of the peat swamp forest used in restoration were *Gonystylus bancanus*, *Callophylum lowii*, *Dyera lowii*, *Cratoxylon arborescens*, *Shorea* spp., and *Shorea balangeran*. Total estimated emission reduction was higher in natural forest than in logged-over forest and wind-disturbed forest: 207.36 CO_2 Mg h^{-1}, 161 CO_2 Mg h^{-1}, and 65.87 CO_2 Mg h^{-1}, respectively. This study confirmed that above-ground peat swamp carbon sequestering can be restored even in severely degraded peatlands. Avoiding vegetation loss is an important part of restoration activities and is needed to continuously maintain the below-ground carbon stock in various types of land use. Restoration of biomass requires a longer time, and integrated restoration actions are needed to achieve economic, ecological, and social success. The utilization of local trees in peat swamp forests is a preferred restoration strategy for enhancing carbon stock and reducing the amount of carbon released into the atmosphere. In addition to carbon sequestration, natural regeneration in the peat ecosystem reduces peat decomposition caused by canal drainage.

Peat restoration must be included in the national mitigation strategy to achieve Indonesia's national commitment to reducing its carbon emissions by 29–41% by 2030. This study provides strong evidence that peat restoration is an effective and natural solution for peat-rich countries, particularly Indonesia.

References

Agus F, Runtunuwu E, June T et al (2009) Carbon budget in land use transitions to plantation. Indones J Agric Res Dev 29(4):119–126

Bastoni (2019) Pengembangan Agro-Silvo-Fishery (Wana-Mina-Tani) untuk restorasi gambut berbasis KHG. Pertemuan ilmiah restorasi gambut. Kedeputian Penelitian dan Pengembangan Badan Restorasi Gambut, Jakarta

Basuki I (2017) Carbon stocks and emissions factors of tropical peat swamp forest in response to land cover changes in West Kalimantan, Indonesia. Oregon State University, Corvallis

Binkley CS, Apps MJ, Dixon RK et al (1997) Sequestering carbon in natural forests. Crit Rev Environ Sci Technol 27:S23–S45

Brown S (1997) Estimating biomass and biomass change of tropical forests: a primer. FAO Forestry Paper 134, Rome, p 87

Brown S, Lugo AE (1982) The storage and production of organic matter in tropical forests and their role in the global carbon cycle. Biotropica 14(3):161–187

CCROM IPB (2021) Laporan akhir pendugaan emisi GRK setelah intervensi restorasi gambut. Laporan Internal Kedeputian Penelitian dan Pengembangan Badan Restorasi Gambut, Jakarta

Driessen PM, Rochimah L (1976) Physical properties of lowland Peats from Kalimantan. In: Peat and podzolic soils and their potential for agriculture in Indonesia. Proceedings ATA 106 Midterm Seminar, Tugu, October 1976. Soil Research Institute, Bogor

Giesen W (2004) Introduction. In: Causes of peat swamp forest degradation in Berbak NP, Indonesia, and recommendation for restoration. ARCADIS Euroconsult, Arnhem, pp 11–19

Gunawan H, Kobayashi S, Mizuno K et al (2012) Peat swamp forest types and their regeneration in Giam Siak Kecil-Bukit Batu biosphere reserve, Riau, East Sumatra, Indonesia. Mires Peat 10:5

Hirano T, Jauhiainen J, Inoue T et al (2009) Controls on the carbon balance of tropical peatlands. Ecosystems 12(6):873–887. https://doi.org/10.1007/s10021-008-9209-1

Hiratsuka M, Toma T, Diana R et al (2006) Biomass recovery of naturally regenerated vegetation after the 1998 forest fire in East Kalimantan, Indonesia. JARQ 40(3):277–282. https://doi.org/10.6090/jarq.40.277

Hooijer A, Haasnoot M, van der Vat M et al (2008) Master plan for the conservation and development of the ex-mega rice project area in Central Kalimantan. Technical report no 2, Euroconsult Mott MacDonald/Deltares, Jakarta/Delft

Hooijer A, Page SE, Jauhiainen J (2009) Summary interim report, 2007–2008: first tentative findings on hydrology, water management, carbon emissions and landscape ecology. Delft Hydraulics, Delft

Hooijer A, Page SE, Canadell JG et al (2010) Current and future CO_2 emissions from drained peatlands in Southeast Asia. Biogeosciences 7:1505–1514. https://doi.org/10.5194/bg-7-1505-2010

Hooijer A, Page SE, Jauhiainen J et al (2012) Subsidence and carbon loss in drained tropical peatlands. Biogeosciences 9(3):1053–1071. https://doi.org/10.5194/bg-9-1053-2012

Istomo (2002) Phosphorus and calcium contents in the soil and biomass of peat swamp forest: a case study at the concession area of PT. Diamond Raya Timber, Bagan Siapi-api, Riau Province. Dissertation, Bogor Agricultural University

Masripatin N, Ginoga K, Pari G et al (2010) Carbon stocks on various types of forest and vegetation in Indonesia. Research and Development of Forestry Department, Bogor

Masson-Delmotte V, Zhai P, Pörtner HO et al (eds) (2019) Climate change and land: an IPCC special report on climate change, desertification, land degradation, sustainable land management, food security, and greenhouse gas fluxes in terrestrial ecosystems. IPCC, Geneva

MoEF (Ministry of Environment and Forestry) (2018) Indonesia second biennial update report: under the United Nations framework convention on climate change. Directorate General of Climate Change, Ministry of Environment and Forestry, Jakarta

Murdiyarso D, Herawati H, Iskandar H (2005) Carbon sequestration and sustainable livelihoods: a workshop synthesis. CIFOR, Bogor

Murdiyarso D, Hergoualc'h K, Verchot LV (2010) Opportunities for reducing greenhouse gas emissions in tropical peatlands. PNAS 107(46):19655–19660. https://doi.org/10.1073/pnas.0911966107

Neuzil SG (1997) Onset and rate of peat and carbon accumulation in four domed ombrogenous peat deposits, Indonesia. In: Rieley JO, Page SE (eds) Biodiversity and sustainability of tropical peatlands, Proceedings of the International symposium on biodiversity, environmental importance and sustainability of tropical peat and peatlands, Palangkaraya, September 1995. Samara Publishing, Cardigan, pp 55–72

Novita N, Kauffman JB, Hergoualc'h K et al (2020) Carbon stocks from peat swamp forest and oil palm plantation in Central Kalimantan, Indonesia. In: Djalante R, Jupesta J, Aldrian E (eds) Climate change research, policy and actions in Indonesia: science, adaptation and mitigation. Springer Climate. Springer, Cham, pp 203–227. https://doi.org/10.1007/978-3-030-55536-8_10

Page SE, Rieley JO, Shotyk OW et al (1999) Interdependence of peat and vegetation in a tropical peat swamp forest. Philos Trans R Soc B 354(1391):1885–1897. https://doi.org/10.1098/rstb.1999.0529

Page SE, Wüst RAJ, Weiss D et al (2004) A record of Late Pleistocene and Holocene carbon accumulation and climate change from an equatorial peat bog (Kalimantan, Indonesia): implications for past, present and future carbon dynamics. J Quat Sci 19(7):625–635. https://doi.org/10.1002/jqs.884

Page SE, Hosciło A, Wösten H et al (2008) Restoration ecology of lowland tropical peatlands in Southeast Asia: current knowledge and future research directions. Ecosystems 12(6):888–905. https://doi.org/10.1007/s10021-008-9216-2

Page SE, Rieley JO, Banks JC (2011) Global and regional importance of the tropical peatland carbon pool. Glob Change Biol 17(2):798–818. https://doi.org/10.1111/j.1365-2486.2010.02279.x

Palm C, Tomich T, van Noordwijk M et al (2004) Mitigating GHG emissions in the humid tropics: case studies from the alternatives to Slash-and-Burn Program (ASB). Environ Dev Sustain 6(1–2):145–162. https://doi.org/10.1023/B:ENVI.0000003634.50442.ca

Rieley JO, Page SE, Jauhiainen (eds) (2005) Wise use of tropical peatlands: focus on Southeast Asia. Alterra, Wageningen

Takahashi H, Shimada S, Ibie BF et al (2002) Annual changes of water balance and a drought index in a tropical peat swamp forest of Central Kalimantan, Indonesia. In: Rieley JO, Page SE, Setiadi B (eds) Peatlands for people: natural resource functions and sustainable management. Proceedings of the International symposium on tropical peatlands, Jakarta van der Meer PJ, Verwer CC (2011) Towards a reference carbon value for peat swamp forest in Southeast Asia based on historical inventory data. In: Workshop on tropical wetland ecosystems of Indonesia: science needs to address climate change adaptation and mitigation. Sanur Beach Hotel, Bali

Verwer CC, van der Meer PJ (2010) Carbon pools in tropical peat forest: towards a reference value for forest biomass carbon in relatively undisturbed peat swamp forests in Southeast Asia. Alterra report 2108. Alterra, Wageningen

Wahyunto, Dariah A, Agus F (2010) Distribution, properties, and carbon stock of Indonesian peatland. In: Chen ZS, Agus F (eds) Proceedings of International workshop on evaluation and sustainable management of soil carbon sequestration in Asian countries, Bogor, September 2010. Indonesian Soil Research Institute, Bogor; Food and Fertilizer Technology Center for the Asian and Pacific Region, Taipei. National Institute for Agro-Environmental Sciences, Tsukuba

Yu Z, Loisel J, Brosseau DP et al (2010) Global peatland dynamics since the last glacial maximum. Geophys Res Lett 37(13):L13402. https://doi.org/10.1029/2010GL043584

Part III
Transformation

Chapter 9
Water Management for Integrated Peatland Restoration in Pulau Tebing Tinggi PHU, Riau

Sigit Sutikno, Rinaldi, Muhamad Yusa, Besri Nasrul, Yesi, Chairul, Adhy Prayitno, Akhbar Putra, and Muhammad Gevin Ardi

Abstract Water management is an important aspect for hydrological restoration in the tropical peatland because the availability of water is not evenly distributed in the dry and rainy seasons. The aim of this study was to conduct action research focusing on water management for integrated peatland restoration at Pulau Tebing Tinggi Peatland Hydrological Unit (PHU), Riau, Indonesia. The actions were to implement some research results and findings by developing demo-plots for a pilot project and analyzing their impact. The pilot project for water management was developed at Pulau Tebing Tinggi PHU not only for the purpose of peat rewetting, but also to support revegetation efforts and revitalization of livelihood. Pulau Tebing Tinggi PHU, located in Kepulauan Meranti Regency, Riau Province, is susceptible to peat fires. In 2014, big peat fires occurred in Pulau Tebing Tinggi PHU and several peatland areas in Riau, causing a haze disaster that lasted for about 2 months. The

S. Sutikno (✉) · Rinaldi · M. Yusa
Civil Engineering Department, Center for Disaster Studies, University of Riau, Pekanbaru, Riau, Indonesia
e-mail: sigit.sutikno@lecturer.unri.ac.id

B. Nasrul
Soil Science Department, Center for Disaster Studies, University of Riau, Pekanbaru, Riau, Indonesia

Yesi
Sociology Science Department, Center for Disaster Studies, University of Riau, Pekanbaru, Riau, Indonesia

Chairul
Chemical Engineering Department, Center for Disaster Studies, University of Riau, Pekanbaru, Riau, Indonesia

A. Prayitno
Mechanical Engineering Department, Center for Disaster Studies, University of Riau, Pekanbaru, Riau, Indonesia

A. Putra · M. G. Ardi
Center for Disaster Studies, University of Riau, Pekanbaru, Riau, Indonesia

© The Author(s) 2023
K. Mizuno et al. (eds.), *Vulnerability and Transformation of Indonesian Peatlands*,
Global Environmental Studies, https://doi.org/10.1007/978-981-99-0906-3_9

disaster produced a sickening and deadly cloud of smoky pollution that not only threatened Indonesia but also neighboring countries.

The Thornthwaite-Mather water balance (TMWB) model was applied for water balance analysis as a basis for water management in the research site. A masterplan for water management was developed which was integrated with revegetation and revitalization of livelihood approaches. Canal block constructions, paludiculture, and aquaculture were the integrated activities carried out to support peatland restoration. Two types of canal blocks, whose main materials were wood and vinyl sheet pile, were introduced in this pilot project. Four key parameters of peatland restoration progress were monitored periodically, namely water table, land subsidence, CO_2 emissions, and vegetation growth. This research found that by applying water management properly, the water table can be maintained at a stable and high level in wet peatlands. Water management by applying canal blocking has a good impact for keeping groundwater elevation and keeping peatland in a wet condition for a distance of 400 m upstream from the canal block.

Keywords Water management · Hydrological restoration · Tropical peatland

9.1 Introduction

Peatlands are wetland ecosystems that are characterized by the accumulation of organic matter called "peat," which is derived from dead and decaying plant material under high water saturation conditions (Parish et al. 2008). Peatland ecosystems are the most efficient carbon sinks on the planet because peatland plants capture the CO_2 naturally released from the peat, maintaining a balance. It was found that peatland only covers 3% of the area of the world, but they store about 15–30% of the world's carbon in the form of peat or organic matter (Hugron et al. 2013). Peatlands are a type of wetland that are one of the most valuable ecosystems on Earth. They are critical to conserve global biodiversity, provide safe drinking water, minimize flood risk, and help address climate change. Peatlands are adapted to the extreme conditions of high water and low oxygen content, of toxic elements, and low availability of plant nutrients (Joosten and Clarke 2002). Peatlands have today become a global topic in relation to a raft of environmental issues that include land fires, declining massive ecosystem functions, and increasing carbon emissions (Sabiham et al. 2016). Peatland fires are hard to extinguish because these occur below the surface, and thus can only be put out by rain or artificial rain using weather modification technology (Sandhyavitri et al. 2018; Sutikno et al. 2020b).

Indonesia holds the largest tropical peatland area, comprising approximately 50% of world's total tropical peatlands. The tropical peatland area in Indonesia is estimated at about 14.91 million ha, spread over Sumatra Island, 6.44 million ha (43%); Kalimantan Island, 4.78 million ha (32%); and 3.69 million ha (25%) in Papua Island (Osaki et al. 2016). Recently, most of those peatlands have already been cleared and drained for food crops and cash crops such as palm oil and other plantations. However, large-scale drainage of peatlands for those purposes has often generated major problems of subsidence, fire, flooding, and deterioration in soil quality (Adesiji et al. 2015; Putra et al. 2016; Ritzema 2008). The Indonesian government, through the Peatland and Mangrove Restoration Agency (Badan Restorasi Gambut dan Mangrove, BRGM) and stakeholders, has been continuously undertaking restoration and fire prevention efforts on those degraded peatlands.

A key factor in tropical peatland restoration is always keeping the peat wet (Giesen and Sari 2018). Hence water management is an important aspect for hydrological restoration in tropical peatlands because the availability of water is not evenly distributed in the dry and rainy seasons. The aim of this study was to conduct action research focusing on water management for integrated peatland restoration at Pulau Tebing Tinggi Peatland Hydrological Unit (PHU), Riau, Indonesia. Action research in this study means that the research was followed by some actions to implement research results and findings by developing some demo-plots for a pilot project. The pilot project for water management at Pulau Tebing Tinggi PHU was aimed not just at peat rewetting but also at supporting revegetation and livelihood revitalization. Pulau Tebing Tinggi PHU, located in Kepulauan Meranti Regency, Riau Province, is at high risk from peat fires.

Canal block constructions, paludiculture, and aquaculture were the integrated activities carried out to support the restoration. Two types of canal blocks, made out of wood and vinyl sheet pile, were introduced in this pilot project. To investigate the performance of the canal blocks, some monitoring and measurement activities are underway, i.e., monitoring and measurement of water table elevation and velocity, land subsidence, and vegetation growth surrounding the canal blocks.

9.2 Methodology

9.2.1 Study Area

The study area is in the island of Tebing Tinggi, Kepulauan Meranti Regency, Riau Province, Indonesia. Based on hydrological assessments, this study found three villages on high priority for hydrological restoration at Pulau Tebing Tinggi PHU, Riau Province, namely, the villages of Tanjung Peranap, Lukun, and Tanjung Sari, as shown in Fig. 9.1. Those villages and many peatland areas in Riau suffered big peat fires in 2014, which caused a haze disaster that lasted nearly 2 months and affected not just Indonesia but neighboring countries as well. Figure 9.1 shows the several hotspots in those villages. This disaster produced a sickening and deadly cloud of smoky pollution which not only threatened the nation but also neighboring countries. Therefore, this action research picked two villages, Lukun and Tanjung Sari, as pilot project areas for the water management model in the tropical peatlands. These villages have different sources of livelihood: the villagers of Tanjung Sari mostly cultivate coconut plantations, while Lukun's economy is mostly based on sago and rubber plantations. This difference will certainly affect the water management system for peatland restoration associated with the dominant land uses in both

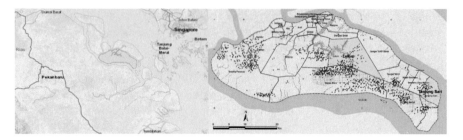

Fig. 9.1 The study area of this action research and the historical fires

villages. Lukun's sago-dominant land use is excellent for a wet environment so there is synergy with the rewetting aspect of peatland restoration. On the other hand, Tanjung Sari, with a coconut-dominant land use, should be cautious with water management because the groundwater elevation should be optimally positioned.

9.2.2 The Peatland of Pulau Tebing Tinggi PHU

Tebing Tinggi Island is dominated by peatlands, whose area is estimated at more than 80% of the island. The peat thickness varies from 1 m to 11 m and forms several peat domes. Based on a field survey, it was found that there are three main peat domes at Pulau Tebing Tinggi PHU as shown in Fig. 9.2 (Nasrul et al. 2020). This study found that the topography of the peat landscape does not reflect peat thickness at Pulau Tebing Tinggi PHU. At this location, high-thickness peat can be found at the edges of the peat dome if it is above a basin or valley in the substratum, while thinner peat can be found in the center if there are mounds in the substratum. Another important finding from this study is that the shape of the peat surface and the substrate surface is not the same; a flat peatland surface may have an undulating substrate and vice versa.

The Ministry of Environment and Forestry has divided the peat domes on the island into three zones, namely, Zone-1, which is located in the Tebing Tinggi Barat Sub-District; and Zone-2 and Zone-3 in the Tebing Tinggi Timur Sub-District, in the middle and east areas, respectively (KLHK 2017). For more detailed hydrological analysis, this study divided Pulau Tebing Tinggi PHU into five sub-PHU which have separate hydrological systems (Fig. 9.3). The sub-PHU division was based on spatial analysis of topographic data, canal networks, and peat depth. Those sub-PHUs are Sub-PHU-1, Sub-PHU-2, Sub-PHU-3, Sub-PHU-4, and Sub-PHU-5, with the area about 41,294 ha (35.8%), 21,840 ha (18.9%), 4230 ha (3.7%), 34,099 ha (29.6%), and 13,864 ha (12%), respectively (Sutikno et al. 2020a).

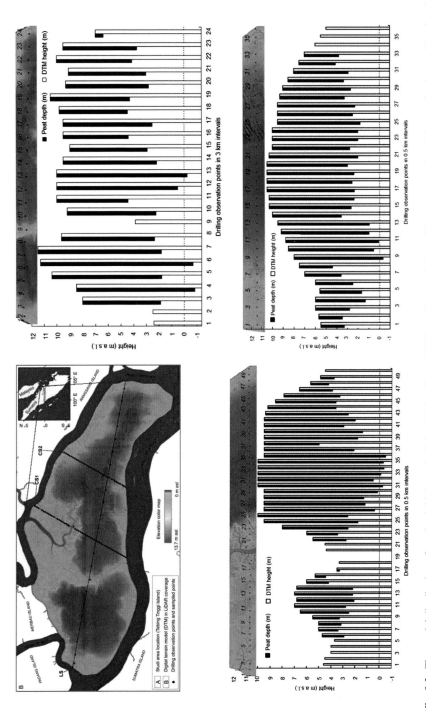

Fig. 9.2 The spatial distribution of peatland thickness in Pulau Tebing Tinggi PHU, and its correlation with the topography of the peat landscape (Reprinted from Nasrul et al. 2020)

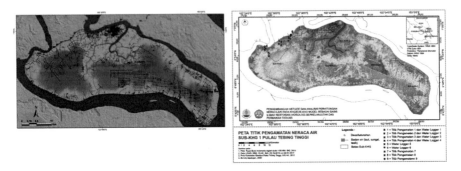

Fig. 9.3 Peat dome zones and the sub-PHU division at Pulau Tebing Tinggi PHU (Gunawan and Budi Triadi 2020)

9.2.3   Peatland Degradation Status

Most of the peatlands in Pulau Tebing Tinggi PHU have been degraded because of the loss of water—which has been directed to plantations—and fires in the last several years. The peatland degradation status in Pulau Tebing Tinggi PHU was analyzed using Landsat Satellite Imagery 8 Operational Land Imager (acquisition date, 26 June 2016), which consists of 11 bands. Figure 9.4 shows the result of the peatland degradation analysis of Pulau Tebing Tinggi PHU. The remaining primary peat swamp forest is only 31% of the original forest area, while the rest of the original forest has been used and degraded as burn scars, disturbed/regrowth peat swamp forest, and agriculture mosaic as shown in Table 9.1 (Sutikno et al. 2020a).

9.2.4   Thornthwaite Water Balance Model for Water Management

This study applied the Thornthwaite-Mather water balance (TMWB) model in Pulau Tebing Tinggi Peatland Hydrological Unit (PHU), Riau Province, Indonesia to analyze the water balance for water management on degraded peatlands. The data required for analysis were monthly rainfall and temperature, land use, degree of peat decomposition, and the latitude of the location. Using the estimated day length based on latitude and month, as well as temperature, the monthly potential evapotranspiration (PET) was estimated. Actual evapotranspiration (ET) will be equal to PET if there is sufficient rainfall (P) and soil moisture, otherwise, Hamon's method is used to estimate ET from PET. When the quantity of P exceeds PET, then the water will enter the soil moisture storage. After the field capacity of the soil is exceeded, the excess becomes runoff. The remaining water in soil moisture storage at the end of the month is carried over to the following month. The general formula of the water balance for the peatland is as follows.

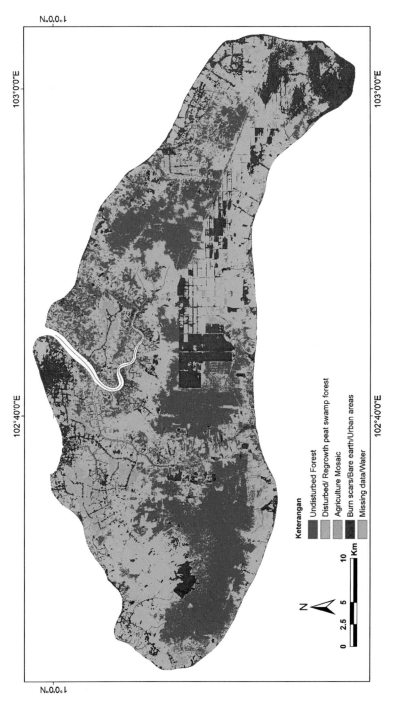

Fig. 9.4 Peatland degradation area in Pulau Tebing Tinggi PHU (Sutikno et al. 2020a)

Table 9.1 The status of peatland degradation in Pulau Tebing Tinggi PHU (Sutikno et al. 2020a)

No	Land cover type	Area (ha)	Percentage (%)
1	Primary peat swamp forest (PSF)	41,378	31
2	Disturbed/regrowth PSF	56,889	43
3	Agriculture mosaic	14,554	11
4	Burn scars	18,091	14
5	Missing data/water	1286	1
Total		132,198	

$$\pm \Delta S = \text{Input} - \text{Output} \tag{9.1}$$

where, ΔS is an addition or reduction to the existing storage in the peat dome.

Analysis of water balance spatially and temporally in peatlands is important to determine the water quantity in spatial and time series. With the water balance analysis, the time (months) and the amount (volume) of surplus or deficit of water in a study area can be understood. The status of "surplus" and "deficit" in a peatland area as a result of water balance analysis is very important to know in order to design the water management program.

9.3 Results and Discussion

9.3.1 The Basic Concept of Water Management

Always keeping the peat wet in tropical degraded peatland requires the proper application of a water management system that is developed based on the water balance characteristics of the study area. Water balance is an estimation of the amount of water in a system based on the hydrologic cycle which consists of water input from precipitation and output from evapotranspiration, outflow, and seepage. The water balance analysis can determine the time (months) and the amount (volume) of surplus and deficit of water in a study area. Thus, it can be calculated whether the excess water can be used to cover the deficit in the dry season. Alternative methods for storing and managing excess water in the rainy season can be applied to keep peat wet in the dry season by keeping the groundwater level (GWL) high.

Figure 9.5 shows the water balance condition of Sub-PHU-1 Pulau Tebing Tinggi in 2019 as results of the TMWB model simulation. It clearly shows that in the dry season, particularly January–February and June–September, that the rainfall was limited. The water balance in these months was a deficit which means the water input was less than the water output. This condition caused the groundwater level to decrease and made the peatland became dry. On the other hand, in the rainy season, particularly in October–December and March–May, the water balance was a surplus which means the water input was more than the water output. The excess water in the

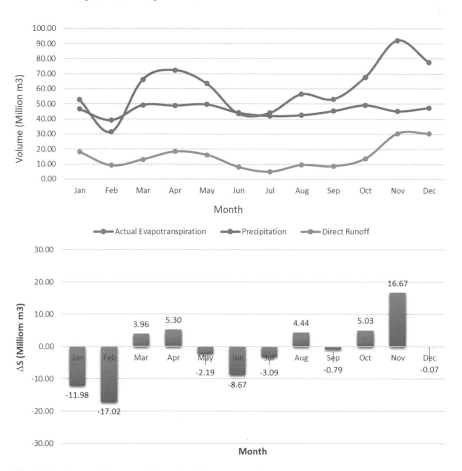

Fig. 9.5 The monthly water balance in Sub-PHU-1 of Pulau Tebing Tinggi (Gunawan and Budi Triadi 2020)

PHU flows into canals and rivers and eventually empties into the sea. However, in order to mitigate the water deficit in the dry season, that excess water should be managed and stored optimally to wet the peatlands in the dry season.

The basic concept of water management in tropical peatlands is to store as much water as possible and for as long as possible in the peatlands while water from the rains (without any flooding) continues to wet the peatlands in the dry season. The water is expected to be stored in the peat dome, which functions as a peatland reservoir, as presented in Fig. 9.6. Water that flows from the peat dome through canals is stopped by a series of blocks along the canal to reduce its velocity and to store it in the canals. The stored water is supposed to infiltrate horizontally to the peatlands so that they are always kept wet. Finally, excess water is drained into the sea to prevent flooding.

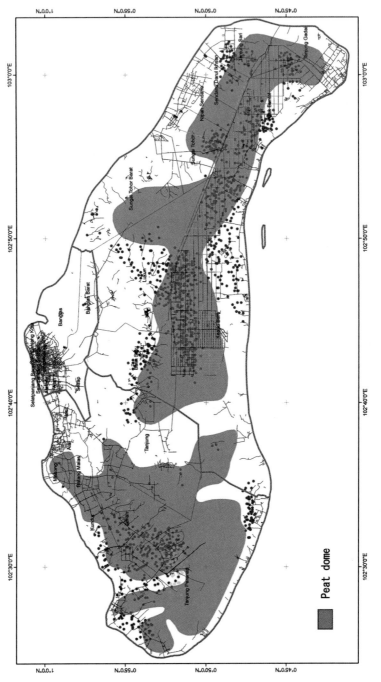

Fig. 9.6 Peat dome as a water reservoir in Pulau Tebing Tinggi PHU (KLHK 2017)

9.3.2   Masterplan for Water Management

A masterplan for water management in Lukun was developed based on a participatory approach together with the local community using spatial data and water balance of the study site. Generally, the masterplan was divided into two zones (Fig. 9.7), i.e., (a) Zone I, where the area is prioritized for peat rewetting and for supporting paludiculture and fishing ponds cultivated by local residents; and (b) Zone II, where the area is prioritized for peat rewetting and fire prevention.

Zone I has been utilized for various economic activities by the local community of Lukun, e.g., developing fishponds in the peat swamp, paludiculture, and other economic activities. Zone II, where much of the fire occurred in 2014, was prioritized for peat rewetting and peatland fire prevention. However, there is a possibility that this zone could be developed for economic activities, such as aquaculture along the canal when the canal block is built.

The masterplan for water management of Tanjung Sari was also developed based on a participatory approach together with the local community. Water management through the construction of a canal block in Tanjung Sari was prioritized to control water allocation for paddy field activities in downstream areas as well as for peat rewetting in upstream areas, as presented in Fig. 9.8. At the moment, paddy fields are being planted in the first phase of planting since these had been developed on the back of water availability. Overall, the masterplan for water management in Tanjung Sari consists of six canal blocks, which are made of vinyl sheet pile, and three areas for paludiculture.

9.3.3   Implementation of Water Management

Peatland restoration efforts in Indonesia involve local communities because these are the main actors in both protecting and damaging peatlands. In addition, peatland restoration efforts are not carried out overnight; rather, it is a continuous process with a long view. As such, local communities play a very important role in the success of peatland restoration efforts. Based on these considerations, the canal blocks for water management should be designed as simply as possible so they could be built by the local community as needed. Also, the material for the canal block model should be locally available. However, the canal block model must function optimally in order to ensure the availability of water to wet the peatlands even in the dry season, to irrigate fishponds and ricefields, and to prevent flooding. For those purposes, the canal block designed for Lukun used wood as shown in Fig. 9.9, and one using vinyl sheet pile was designed for Tanjung Sari as shown in Fig. 9.10.

The canal block model in Lukun uses locally sourced wood that is lightweight, easily available, and inexpensive. The model as designed and as built is presented in Fig. 9.9. The model is equipped with a sluice gate for flexibility and to maintain water levels in the canal during both the rainy and dry seasons. In the rainy season,

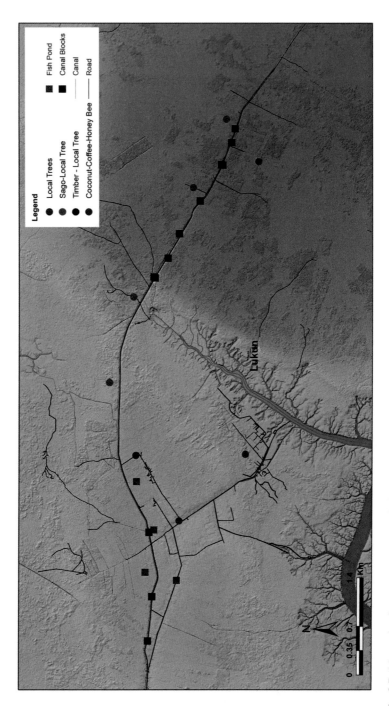

Fig. 9.7 Masterplan for water management in Lukun

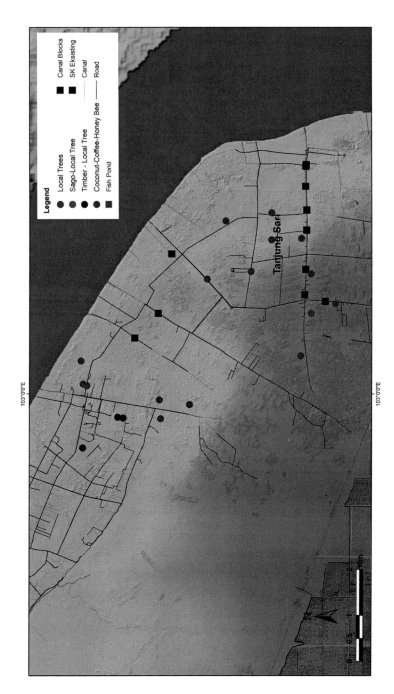

Fig. 9.8 Masterplan for water management in Tanjung Sari

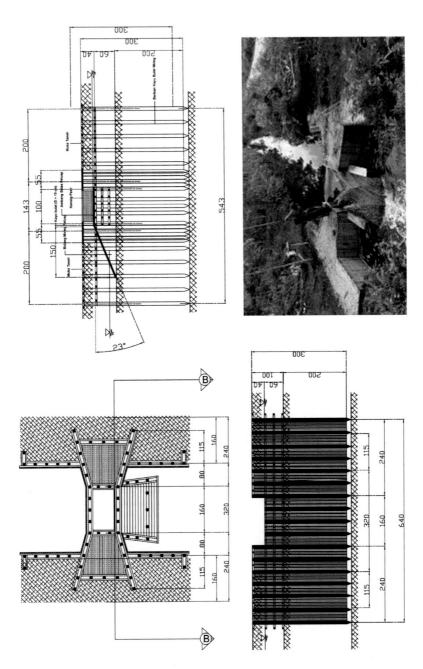

Fig. 9.9 Canal block model for water management in Lukun

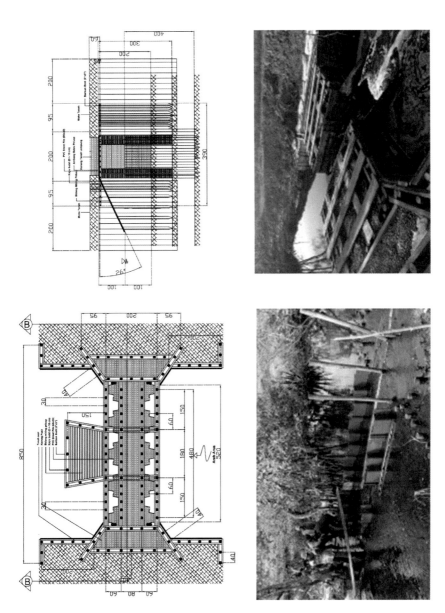

Fig. 9.10 Canal block model for water management in Tanjung Sari

the canal blocks are opened to prevent flooding, and in the dry season the canal blocks are closed to keep the peatlands wet. The canal block is also built with a spillway to prevent turbulence that could cause erosion and affect its structural stability. The local community has adopted the masterplan, building seven canal blocks since 2017.

The canal block model in Tanjung Sari uses vinyl sheet pile, as shown in Fig. 9.10. This material replaces wood, which is quite limited in the area, and works best with the high tide. The vinyl sheet pile has two layers, with the option for a filler material for greater stability. It is also equipped with a sluice gate for flexibility and to maintain water levels in the canal during both the rainy and dry seasons. This canal block model has been constructed in Tanjung Sari.

9.3.4 Monitoring the Progress of Restoration Efforts

Efforts to restore peatlands to their near-natural conditions require a long-term process. Therefore, the progress of the peatland restoration should be monitored periodically. We applied four methods to monitor the four key parameters of the peatland restoration efforts, namely, water table, land subsidence, CO_2 emissions, and vegetation growth. The water table has been monitored periodically using SP-MATRIKLAGA (Monitoring Dipwell for Groundwater Table and Fire Risk at Peatland), equipment developed by the University of Riau which uses water loggers. The equipment also uses a floating system that can determine the groundwater depth and the level of fire risk directly from the instrument as presented in Fig. 9.11a. Through this equipment, the impact of canal blocking on rewetting in the peatland can be analyzed. Land subsidence is monitored periodically every year using a fixed reference fix pole as shown in Fig. 9.11b. Nine subsidence pole were installed in Pulau Tebing Tinggi PHU, which are monitored every year. The CO_2 emissions are measured with an environmental gas monitor (EGM) chamber in a closed dynamic chamber as shown in Fig. 9.11c, and vegetation growth is measured periodically using an Unmanned Aerial Vehicle (UAV) as shown in Fig. 9.11d.

9.3.5 Impact of Water Management

This section presents the impact of implementing the masterplan for water management in the tropical peatlands. To analyze the impact, eight dipwells were made to monitor groundwater fluctuations with the distance of 1 m, 101 m, and 201 m from the canal for each of the three transects as presented in Fig. 9.12. The Canal Block-1 is located about 100 m downstream of the Transect-1, and about 200 m upstream of the Transect-2. The Canal Block-2 is located about 100 m downstream of the Transect-2, and about 100 m upstream of the Transect-3.

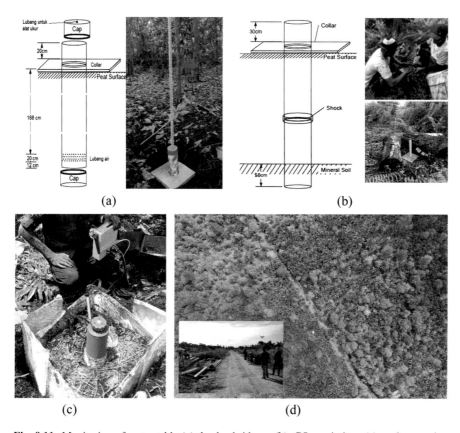

Fig. 9.11 Monitoring of water table (**a**), land subsidence (**b**), CO_2 emissions (**c**), and vegetation growth (**d**)

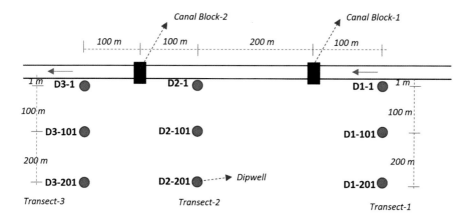

Fig. 9.12 The setting of dipwells for monitoring the impact of water management in the research site

Figure 9.13 shows the daily groundwater level monitoring results before and after water management with canal blocking. The Canal Block-1, which is on the upstream side, was built in mid-November 2018, and the Canal Block-2 was built in early November 2018. When the Canal Block-1 was constructed, the groundwater level at dipwell D1-1, which is about 100 m upstream of the Canal Block-1, increased significantly, but the other two wells did not. This means that canal blocking only affects upstream areas. Groundwater levels at dipwell D1-1 were stable at an elevation of about 8.5 m. This means that the water management with canal blocking worked well to maintain the elevation of the groundwater table and to keep the peatland in a wet condition.

When the Canal Block-2 was constructed, there was an effect on groundwater levels in dipwell D2-1 which is about 100 m upstream. However, this blocking had no impact on dipwell D1-1 which is about 400 m upstream. This means that in this case the canal blocking was successful in wetting an area of no more than 400 m upstream. Groundwater level fluctuations downstream of the canal block were not affected by canal blocking in the upstream part. In the rainy season from October to December 2018, the groundwater level fluctuated dynamically according to rainfall occurrence. In the dry season from January to March 2019, the groundwater level remained stable. There were very few rains, but GWL on dipwell D1-1 and D2-1 was still high because of canal blocking. There was a slight groundwater level decrease in February due to evapotranspiration, but the water table remained high to keep the peatland in the wet condition although in the dry season. It means that the water management by canal blocking had a significant impact on increasing the groundwater levels which can reduce the risk of peatland fire.

To analyze the impact of canal blocking on wetting efforts on peatlands, groundwater level fluctuations were analyzed from three dipwells on each transect with a perpendicular distance of 1 m, 101 m, and 201 m, respectively. Figures 9.14 and 9.15 show the relationship between the increase of water levels in the canal due to canal blocking and the increase of GWL in peatland with a distance of 101 m and 201 m to the canal on Transect-1 and Transect-2 respectively. The increase of GWL in peatland due to canal blocking varied in each location depending on the distance to the canal. The canal blocking does not have an impact on rewetting at a distance of 101 m and 201 m to the canal if the increase in water levels at the canal is less than 0.4 m and 0.45 m in case of Transect-1 (Fig. 9.14) and Transect-2 (Fig. 9.15) respectively. Water levels in the canal with an increase of less than 0.4 m on Transect-1 and 0.45 m on Transec-2 due to rainfall did not affect the increase in groundwater levels in the peatland. The groundwater level in peatlands in these conditions has decreased which is probably due to higher evapotranspiration than rain. The increase of GWL due to canal blocking varied in each location depending on the distance to the canal. The increase of water level at the canal of about 0.58 m on Transect-1 caused the increase of groundwater level of about 0.42 m and 0.34 m for a distance of 101 m and 201 from the canal respectively (Fig. 9.14). The increase of water level at the canal of about 0.57 m on Transect-2 caused the increase of groundwater level of about 0.32 m and 0.30 m for a distance of 101 m and 201 from the canal respectively (Fig. 9.15).

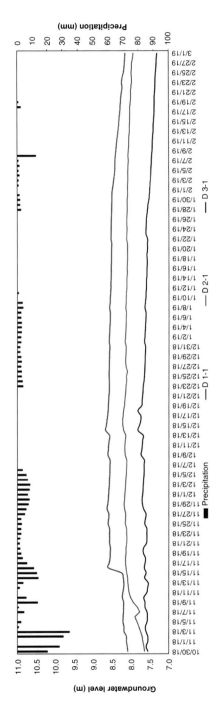

Fig. 9.13 Daily water table fluctuation because of water management of three dipwells along the canal with 1 m distance from the canal, from October 2018 to March 2019

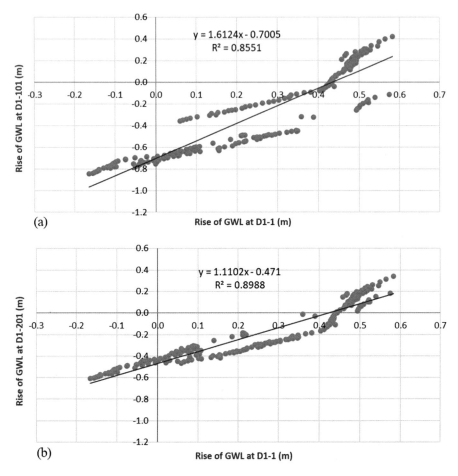

Fig. 9.14 The relationship between the increase in water levels at the canal (D1-1) and the increase of GWL in peatlands with the distance of 100 m (**a**) and 200 m (**b**) to the canal for Transect-1

In comparison with GWL fluctuation because of canal blocking, Fig. 9.16 presents the increase of GWL on the peatlands with a distance of 101 m to the canals without the impact of canal blocking on Transect-3. It shows that there is an increase of GWL on peatlands in line with the increase in the water level in the canal. However, the increase in water level in canals tends to be smaller than the increase of GWL in the peatlands. This shows that water fluctuations in canals and GWL on peatlands are caused by precipitation, without the impact of canal blocking. The increase in water level in the canals is smaller than the increase of GWL in peatlands due to the large water loss due to the absence of canal blocking.

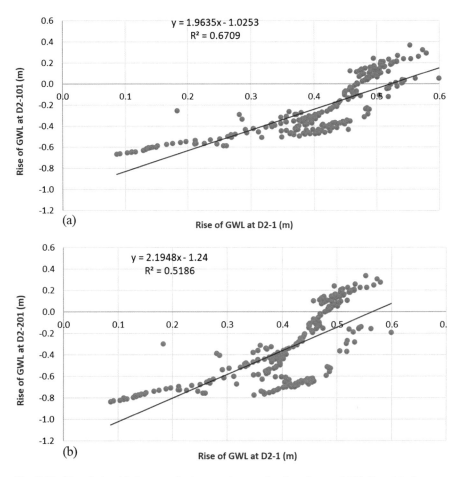

Fig. 9.15 The relationship between the increase in water levels at the canal (D1-1) and the increase of GWL in peatlands with the distance of 100 m (**a**) and 200 m (**b**) to the canal for Transect-2

9.4 Conclusion

This study presented a pilot project for water management not only with the purpose of peat rewetting but also for supporting revegetation efforts and revitalization of livelihood in Pulau Tebing Tinggi Peatland Hydrological Unit (PHU). Two masterplans of water management for both Lukun and Tanjung Sari were developed with the participation of the respective communities. The latter have adopted the masterplans by constructing seven wooden canal blocks in Lukun and a sheet pile canal block in Tanjung Sari. These canal blocks were integrated with other restoration activities, such as paludiculture and aquaculture. This research found that by applying water management properly, the water table can be maintained at a stable

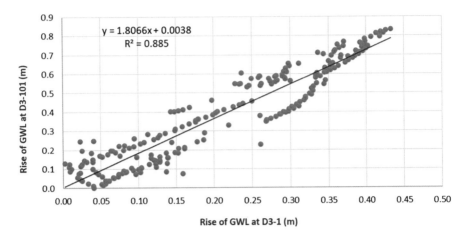

Fig. 9.16 The relationship between the increase in water levels at the canal (D1-1) and the increase of GWL in peatlands with the distance of 100 m to the canal for Transect-3

and high level to wet the peatlands. Water management by applying canal blocking worked well to keep groundwater elevation and keep peatland in a wet condition for an area 400 m upstream from the canal block. Canal blocking will have a good impact on wetting in peatlands up to a distance of 201 m perpendicular to the canal if the water level rise in the canal is more than 0.45 m.

Acknowledgments The authors would like to thank to Peatland and Mangrove Restoration Agency (BRGM), The Republic of Indonesia, for its support and facilitation of this research.

References

Adesiji AR, Mohammed TA, Nik Daud NN et al (2015) Impacts of land-use change on peatland degradation: a review. Ethiop J Environ Stud Manag 8(2):225–234. https://doi.org/10.4314/ejesm.v8i2.11

Giesen W, Sari ENN (2018) Tropical peatland restoration report: the Indonesian case. https://doi.org/10.13140/RG.2.2.30049.40808

Gunawan H, Budi Triadi L (eds) (2020) Neraca air: kesatuan hidrologis gambut. Badan Restorasi Gambut, Republik Indonesia, Jakarta

Hugron S, Bussières J, Rochefort L (2013) Tree plantations within the context of ecological restoration of peatlands: a practical guide. Peatland Ecology Research Group, Université Laval, Québec

Joosten H, Clarke D (2002) Wise use of mires and peatlands: background and principles including a framework for decision-making. International Mire Conservation Group, Greifswald; International Peat Society, Jyväskylä

KLHK (Kementerian Lingkungan Hidup dan Kehutanan) (2017) Penetapan peta fungsi ekosistem gambut skala 1: 50.000 pada KHG Pulau Bengkalis, KHG Pulau Tebing Tinggi, KHG Sungai Kampar–Sungai Gaung, KHG Sungai Gaung–Sungai Batang Tuaka, KHG Sungai Kapuas–

Sungai Terentang. In: Keputusan Menteri Lingkungan Hidup Dan Kehutanan Republik Indonesia Nomor: SK.295/Menlhk/Setjen/PKL.0/6/2017

Nasrul B, Maas A, Utami SNH et al (2020) The relationship between surface topography and peat thickness on Tebing Tinggi Island, Indonesia. Mires Peat 26:18. https://doi.org/10.19189/MaP. 2019.OMB.StA.1811

Osaki M, Nursyamsi D, Noor M et al (2016) Peatland in Indonesia. In: Osaki M, Tsuji N (eds) Tropical peatland ecosystems. Springer, Tokyo, pp 49–58. https://doi.org/10.1007/978-4-431-55681-7_3

Parish F, Sirin A, Charman D et al (eds) (2008) Assessment on peatlands, biodiversity and climate change: main report. Global Environment Centre, Kuala Lumpur; Wetlands International, Wageningen

Putra EI, Cochrane MA, Vetrita Y et al (2016) Degraded peatlands, ground water level and severe peat fire occurrences. In: 15th International paet congress 2016., Kuching, 15–19 August 2016, pp 327–331

Ritzema HP (2008) The role of drainage in the wise use of tropical peatlands. In: Wösten JHM, Rieley JO, Page SE (eds) Restoration of tropical peatlands. Alterra, Wageningen, pp 78–87

Sabiham S, Winarna, Pulunggono HB et al (2016) What is the way forward on Indonesian Peatland? In: 15th International peat congress 2016, Kuching, 15–19 August, pp 39–45

Sandhyavitri A, Perdana MA, Sutikno S et al (2018) The roles of weather modification technology in mitigation of the peat fires during a period of dry season in Bengkalis, Indonesia. IOP Conf: Mater Sci Eng 309(1):012016. https://doi.org/10.1088/1757-899X/309/1/012016

Sutikno S, Hidayati N, Rinaldi et al (2020a) Classification of tropical peatland degradation using remote sensing and GIS technique. AIP Conf Proc 2255:070022. https://doi.org/10.1063/5.0013881

Sutikno S, Amalia IR, Sandhyavitri A et al (2020b) Application of weather modification technology for peatlands fires mitigation in Riau, Indonesia. AIP Conf Proc 2227:030007. https://doi.org/10.1063/5.0002137

Chapter 10
Genetic Diversity in Peatland Restoration: A Case of Jelutung

Hesti Lestari Tata

Abstract The target of the Indonesian government was to restore at least 2 million ha of degraded peatlands by 2020. This can be achieved by applying three approaches: rewetting, revegetation, and revitalization of the community. In revegetation, the use of native tree species is recommended in the technical guidelines for peatland restoration. The interest in using native plant species for ecosystem restoration, particularly for peatland restoration, is increasing significantly. The native species may adapt well to the environment and usually correspond with the people's preference. An important concern in revegetation is the selection of suitable genetics of the planting stocks. In peatland restoration, a seed source with high genetic diversity should be used since it maintains the genetic entity. On the other hand, low genetic diversity used for peatland restoration may result in the genetic drift of the populations. A native tree species, *Dyera polyphylla* (locally known as jelutung), has been widely used for peatland restoration in Indonesia. Many other species can also be developed and require further research. Several community-based seedling nurseries have been established in Sumatra and Kalimantan. The challenge is to emphasize the importance of genetic diversity to the farmers with businesses in a seedling nursery. The necessary strategy of seed sourcing in peatland restoration includes composite provenancing and admixture provenancing.

Keywords Enrichment planting · Native species · Revegetation · Survival

10.1 Introduction

Peatlands cover a large area of Indonesia—around 14.9 million ha (Ritung et al. 2011). After the catastrophic fires in 2015, the Indonesian government targeted the restoration of at least 2 million ha of burnt peatlands between 2016 and 2020 (The Presidential Instruction—Inpres No. 1/2016). A nonstructural agency was

H. L. Tata (✉)
Research Center of Ecology and Ethnobiology, National Research and Innovation Agency, Bogor, Indonesia
e-mail: made.hesti.lestari.tata@brin.go.id

© The Author(s) 2023
K. Mizuno et al. (eds.), *Vulnerability and Transformation of Indonesian Peatlands*, Global Environmental Studies, https://doi.org/10.1007/978-981-99-0906-3_10

developed, namely, the Peatland Restoration Agency (Badan Restorasi Gambut, BRG)[1] which was mandated to plan and implement the restoration of the degraded peat ecosystem. This demonstrates the government's commitment to restore degraded peatlands and mangroves and combat peat fires and the haze hazard.

Peatland restoration aims to restore degraded burnt peatlands by applying three approaches: rewetting, revegetation, and revitalization of the community. Rewetting is an effort to rewet the drained peatland by increasing the water content of the peat and raising the ground water level. Revegetation includes wide scope for rehabilitation and replanting the degraded peatlands to improve land cover and economic benefits (BRG 2017a). Page et al. (2009) suggested three similar actions in addition to restoration of the carbon sink and reduction of greenhouse gas emissions. In fact, maintaining a high water table through rewetting and increasing carbon stocks through revegetation can be used to achieve the target of emission reduction (Ritzema et al. 2014; Jauhiainen et al. 2016; Wilson et al. 2016).

The regulation signed by the Minister of Environment and Forestry (number P.16 in 2017) on technical guidelines for the restoration of the peatland ecosystem states that the number of trees planted and have survived in the third year after planting should be at least 500 trees per ha^{-1}. The Minister formally stated that the survival rate of revegetation in peatland restoration should be 90% (Ministry of Environment and Forestry 2018). In a comprehensive review of peatland restoration, Dohong et al. (2018) stated that the early survival rate of revegetation in the Mega Rice Project of Central Kalimantan Province varied between 65% and 85%. They highlighted the need to formulate future revegetation efforts with a strategy appropriate to peat forests. In addition, Page et al. (2009) asserted that the obstacles in revegetation include finding viable planting stocks that are suitable for temporarily high water-logged and dry peatlands. The authors suggested maintaining the ongoing vegetation monitoring in restoration, investigating ecological barriers to tree species regenerating in degraded peatland, and gaining a better understanding of hydrological tolerance in times of drought and in the wet season. Neither article, however, discussed genetic diversity and genetic sources as the criteria for species used in restoration.

Landscape restoration aims to improve the genetic diversity of plant populations to recover the characteristics of species composition and diversity (Brancalion et al. 2011). Restoring a landscape with multifunctionality requires multiple species. Focusing on the functional diversity of tree species is more important than species per se (Aerts and Honnay 2011). In a degraded and burned peatland landscape, the mother trees are not available and genetic materials are disrupted. Therefore, obtaining viable seeds and seedlings is the main constraint (Brancalion et al. 2011).

In peatland landscape restoration, diversity at the gene level is rarely observed. It is necessary to take into account the genetic variation of a species and the variety of plant species planted (Bischoff et al. 2010; Thomas et al. 2014a, b). Genetic diversity

---

[1] It transformed into the Peatland and Mangrove Restoration Agency (Badan Restorasi Gambut dan Mangrove, BRGM) in 2020.

is positively related to the vigor of the tree population, its flexibility, and its ecosystem functions (Reynolds et al. 2012; Thomas et al. 2014a, b). High genetic diversity of plant populations is a fundamental factor in ensuring the long-term success of restoration (Brancalion et al. 2011). The failure of revegetation in peatland restoration is may be caused by low genetic diversity within a species (Thomas et al. 2015).

There are many native peatland tree species; however, information on genetic diversity is very limited. Some reports have shown the genetic diversity of three important tree species that those are naturally grown on peatlands, namely, *Gonystylus bancanus* (Ogden et al. 2008; Widyatmoko and Aprianto 2013), *Shorea macrophylla* (Kanzaki et al. 1996), and *Dyera polyphylla* (Wahyudiningsih et al. 2014; Tata et al. 2018). *D. polyphylla* is widely used in peatland restoration in Indonesia (Tata and Susmianto 2016; Tata et al. 2016). Many other tree species used in peatland restoration merit further study.

This study, which was based on literature review and field observation, aims to highlight the importance of genetic diversity. In addition, we discuss some community-based seedling nurseries of native species in Jambi and Central Kalimantan that provide planting stocks for peatland restoration. The nurseries of native peatland species are usually established by smallholder farmers, who usually collect seeds from wild and planted populations, which will affect the genetic variation of the seedlings.

10.2 Planting Stocks for Peatland Restoration

Following the destructive peat fires in 2015, a series of investigations were conducted in the peat swamp forests in Kerumutan in Riau Province, Kepayang in South Sumatra, and Tumbang Nusa in Central Kalimantan. There is no doubt that many big trees were destroyed in the fires. Figure 10.1a shows the state of the Kepayang peat swamp forest in South Sumatra after the fire. Only a small patch of forest area survived, in which some mother trees were still alive. A year after the fire, some tree species (such as the Euphorbiaceae family) were able to sprout and recover quickly, obviously having suffered a low impact from the fire (Fig. 10.1b). However, the physiology of the threes affected by the fires has not been determined.

Peatland degradation has been characterized based on the change in peatland components, for example, plants, water, and peat, from minimal up to maximal level. These characteristics determine the restoration strategy. Restorability is decreasing in contrast with the degradation level (Joosten 2016). The peatland restoration action is determined by multiple factors, such as level of degradation, location of the peatland in the peatland hydrological unit, and its area status. Although in the regulation, the BRG was mandated to restore million ha of degraded peatland, after reassessment and remeasurement were carried out, the peatland restoration target in Indonesia increased to 2.49 million ha in seven provinces (BRG 2017b). The area and peatland distribution to be restored are shown in Fig. 10.2.

Fig. 10.1 (**a**) Severely burnt peat swamp forest in Kepayang, South Sumatra; (**b**) Sprouting tree a year after the fire

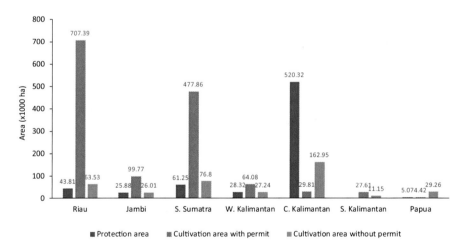

Fig. 10.2 Restoration target of BRG in seven provinces (Source of data: BRG 2017b)

The BRG has classified the target of the three actions in all the target areas. This study focuses on revegetation, which consists of full revegetation, enrichment planting, and natural regeneration. Revegetation is determined by the level of peatland degradation (BRG 2017a).

1. Natural regeneration is applied when the level of degradation is low, and the canopy cover is >50%. The peatland condition allows the recolonization of

Table 10.1   Areas for revegetation and responsible stakeholders (Source of data: BRG 2017a)

Stakeholders	Number of PHU	Revegetation (ha)	Enrichment planting (ha)	Natural regeneration (ha)
Forest concessioners	26	43,912	85,366	41,916
Conservation agency (KSDAE)	8	24,242	58,325	5475
FMU/Forest service	28	65,863	136,572	13,735
Local government (Pemda)	32	12,762	55,024	1259
Total	94	146,779	335,287	62,385

PHU peatland hydrological unit

  native forest species, recruitment of indigenous species, and seed dispersal mechanisms.
2. Enrichment planting is carried out when the level of degradation is moderate, and the canopy cover is between 25% and 50%. In this condition, unassisted regeneration may occur; enrichment planting is recommended to accelerate vegetation cover and to enrich tree diversity.
3. Revegetation (100% planting) is used when the level of degradation is high, and the canopy cover is <25%. Assisted regeneration with native tree species is necessary.

The BRG has calculated the target area for revegetation in seven provinces of Indonesia. The four important stakeholders responsible for revegetation are shown in Table 10.1.

The degraded peatland area to be revegetated is 146,779 ha, and the area for enrichment planting is 335,287 ha. According to the regulation, revegetation should have at least 500 trees/ha that have survived in the third year after planting. The assumption is that enrichment planting only needs 50% trees to be planted in the area for revegetation. If we use a survival rate of 90% yearly, a total of 617 viable seedlings/ha are needed. Therefore, the total number of viable planting stocks (either seedlings or cuttings) required for revegetation is about 90.56 million, and for enrichment planting, it is 103.44 million (in total, 194 million), which is a large number of planting stocks. The intriguing question then becomes: How might we provide the required number of viable planting stocks? Large areas of peat swamp forest have been burnt and degraded, and the standing stocks or mother trees in the natural population are scarce and scattered in patchy areas.

10.3 Avoiding Failure in Revegetation in Peatland Restoration

There is increasing awareness of the importance of planting native tree species in certain ecosystems for restoration. The native species are already adapted to the environment, which improves the success of restoration efforts and thus ecosystem functions (Bischoff et al. 2010; Kettenring et al. 2014). However, the origin of the species alone is not enough. The native species used in restoration should have high genetic diversity and the appropriate germ plasm, as this significantly increases the survival rate from the first year and over time (Broadhurst et al. 2008; Bischoff et al. 2010; Thomas et al. 2015). It has been found that small increase in the genetic diversity of species used in restoration positively enhanced ecosystem services, such as habitat quality, primary productivity, and nutrient retention (Reynolds et al. 2012).

On the other hand, using low-quality materials in revegetation results in high mortality, poor growth, susceptibility to environmental condition, and low reproductive success. These failures may not be evident in the early growth, but become apparent as the trees mature (Bischoff et al. 2010; Thomas et al. 2015). Self-pollination—mating with the offspring—can considerably affect survival and thus lead to genetic drift (Bischoff et al. 2010).

The revegetation project in the peat forest reserve (Hutan Lindung Gambut, HLG) of Bram Itam in Jambi used *D. polyphylla* of which the origin was unknown. However, the genetic analysis revealed that the seedlings were likely brought from Central Kalimantan. The diversity of *D. polyphylla* planted in the HLG of Bram Itam was relatively high ($H' = 0.35$) (Tata et al. 2018). The growth of planted *D. polyphylla* was good: at 7 years after planting, the stem diameter at the breast height had reached 11.2 cm (Tata et al. 2016). This supports the statement of Reynolds et al. (2012) and Thomas et al. (2015).

Wild populations of particular species are usually used as seed sources for forest and landscape restoration. A population in a fragmented and isolated ecosystem may have low genetic diversity owing to limited gene flow. However, in the multifunctionality of landscape, such as agroforestry systems, the gene flow could be increased (Thomas et al. 2015; Tata et al. 2018).

10.4 Sourcing Germ Plasm for Peatland Restoration

The principle in sourcing germ plasm for peatland restoration is using high quality and genetically diverse seeds to maximize the adaptive potential of restoration efforts. Using local seeds may lead to unsuccessful restoration (Broadhurst et al. 2008) for a number of reasons, for example, the low genetic diversity of the local seeds, which come from a fragmented and isolated population. The isolated habitat may lead to self-pollination, which reduces genetic variation. Tata et al. (2018)

Fig. 10.3 The habitat of a wild population of swamp jelutung (*Dyera polyphylla*). Both populations are used as a seed source for peatland restoration in Southern Sumatra. (**a**) Rawasari, Tanjung Jabung Timur, (**b**) Senyerang, Tanjung Jabung Barat, Jambi

reported that several wild populations of *D. polyphylla* from Senyerang and Sungai Aur in Jambi and Tumbang Nusa in Central Kalimantan maintain high genetic diversity since their habitats were surrounded by a complex system of agroforests. An isolated and degraded population of *D. polyphylla*, such as the wild population in Rawasari Jambi, has lower genetic diversity (Tata et al. 2018). Habitat fragmentation may influence pollinators and their behavior, which leads to reduced outcrossing and higher inbreeding (Broadhurst et al. 2008). Wild populations of *D. polyphylla* in Senyerang and Rawasari in Jambi are shown in Fig. 10.3.

10.5   The Community-Based Nursery

Several community-based nurseries (Kebun Bibit Rakyat, KBR) especially for peatland native species have been established in various villages in Sumatra and Central Kalimantan to provide planting stocks for peatland restoration and to encourage farmers to plant native species in peatlands (Fig. 10.4).

In Senyerang village (Jambi), the KBR was established by a forest farmer group (Kelompok Tani Hutan) called Rimba Lestari (Fig. 10.4a) and led by Saman. Some jelutung trees have been certified as mother trees of jelutung by the Forest Seed Technology Office (Balai Teknologi Perbenihan Hutan) (Tata et al. 2016). There are other native tree species in the patchy forest of Senyerang that can be developed as a seed source for peatland restoration, such as medang labu (*Endospermum diadenum*), asam kandis (*Garcinia* sp.), meranti batu (*Shorea uliginosa*), gemor

Fig. 10.4 Community-based seedling nurseries in Jambi and Central Kalimantan: (**a**) Rimba Lestari forest farmer group; (**b**) Jelutung seedlings in a community-based nursery (KBR); (**c**) jelutung planted in Tanjung Jabung Barat; (**d**) jelutung nursery in Tumbang Nusa; and (**e**) nursery jelutung in Hampangen Central Kalimantan

(*Alseodaphne* sp.), and gaharu buaya (*Gonystylus* sp.). However, only three species, namely, asam kandis, jelutung, and medang labu flower yearly.

The KBR in Rawasari village (Jambi) was established and led by Lukman. Jelutung is the only species developed in the KBR (Fig. 10.4b). Other potential species, such as swamp pulai (*Alstonia pneumatophora*), could also be developed. In a field observation, there were some native species of peat swamp forest remains. A big canal was built in the area, which affected the hydrological and ecological conditions.

In Tumbang Nusa in Central Kalimantan, a seedling nursery of native species has become a business and a source of livelihood for the villagers (Fig. 10.4d, e). Four villagers living on the main road developed a seedling nursery and have been selling planting stocks (seedlings and wildings). Various native species are available, including jelutung (as the commonest seedlings), swamp pulai, gelam (*Melaleuca leucodendra*), petai (*Parkia speciosa*), kapur naga (*Calophyllum sclerophyllum*), and jangkang (*Syzygium* sp.). The villagers developed the nursery based on their own initiative, without any cooperation with a farmer group. Figure 10.4c shows a seedling vendor from Tumbang Nusa village. The seedling nursery was developed in an area of 1.4 ha. There were 75,000 seedlings (from various tree species) per ha. The mean price of one seedling was Rp3,000, and the gross revenue was Rp225 million (Sumarhani and Tata 2018).

In providing planting stock, community-based seedling nurseries could enhance peatland restoration in terms of the quantity of seedlings. As mentioned earlier, local species do not necessarily improve the quality of peatland restoration. It has been found that the use of 10–50 selected from healthy and viable individual trees per population and up to 50 populations per species is adequate (Broadhurst et al. 2008). The suitable sourcing strategies include compound provenancing, that is, collecting a mixture of seeds that try to follow gen-flow dynamics; and admixture provenancing, which involves collecting seeds from large populations and various environments, and mixing them before sowing. This generates new populations with a variety of genotypes of wide provenance (Broadhurst and Boshier 2014).

10.6   Conclusion

To achieve the target area of peatland restoration with a survival rate of 90%, there must be a sufficient quantity of high-quality planting stocks to ensure genetic value and plant health. The strategy should include combining seeds not only from local species but also from various environments. As reported by Tata et al. (2018), the planted jelutung in Tanjabar in Jambi and Tumbang Nusa in Central Kalimantan showed no loss of genetic diversity. It was used as a seed source for the recent peatland restoration in the two provinces. Villagers with a seedling nursery business should be made aware of the potential risk of genetic drift if they are collecting seeds from a few individual trees and a limited population. The correct strategy for seed sourcing should be explained to farmers to make them aware of the importance of genetic diversity in peatland landscape restoration.

Acknowledgments   This paper is based on previous series studies and publications in Sumatra and Central Kalimantan. It was presented at a Joint Symposium of Badan Restorasi Gambut and JICA on February 23, 2018, in Jakarta.

References

Aerts R, Honnay O (2011) Forest restoration, biodiversity and ecosystem functioning. BMC Ecol 11:29. https://doi.org/10.1186/1472-6785-11-29

Bischoff A, Steinger T, Müller-Schärer H (2010) The importance of plant provenance and genotypic diversity of seed material used for ecological restoration. Restor Ecol 18(3):338–348. https://doi.org/10.1111/j.1526-100X.2008.00454.x

Brancalion PHS, Viani RAG, Aronson J et al (2011) Improving planting stocks for the Brazilian Atlantic forest restoration through community-based seed harvesting strategies. Restor Ecol 20(6):704–711. https://doi.org/10.1111/j.1526-100X.2011.00839.x

BRG (Badan Restorasi Gambut) (2017a) Rencana kontijensi badan restorasi gambut: perubahan. Badan Restorasi Gambut, Jakarta

BRG (2017b) Rencana strategis badan restorasi gambut tahun 2016–2020. Badan Restorasi Gambut, Jakarta

Broadhurst L, Boshier D (2014) Seed provenance for restoration and management: conserving evolutionary potential and utility. In: Bozzano M, Jalonen R, Thomas E et al (eds) Genetic considerations in ecosystem restoration using native tree species. The state of the world's forest genetic resources: thematic study. Food and Agriculture Organization of the United Nations, Rome.; Bioversity International, Rome, pp 27–37

Broadhurst L, Lowe A, Coates DJ et al (2008) Seed supply for broadscale restoration: maximizing evolutionary potential. Evol Appl 1:587–597. https://doi.org/10.1111/j.1752-4571.2008.00045.x

Dohong A, Aziz AA, Dargusch P (2018) A review of techniques for effective tropical peatland restoration. Wetlands 38:275–292. https://doi.org/10.1007/s13157-018-1017-6

Jauhiainen J, Page SE, Vasander H (2016) Greenhouse gas dynamics in degraded and restored tropical peatlands. Mire Peat 17:06. https://doi.org/10.19189/MaP.2016.OMB.229

Joosten H (2016) Peatlands across the globe. In: Bonn A, Allot T, Evans M et al (eds) Peatland restoration and ecosystem services: science, policy and practice. Cambridge University Press, Cambridge, pp 19–43. https://doi.org/10.1017/CBO9781139177788.003

Kanzaki M, Watanabe M, Kuwahara J et al (1996) Genetic structure of *Shorea macrophylla* in Sarawak, Malaysia. Tropics 6(1/2):153–160. https://doi.org/10.3759/tropics.6.153

Kettenring KM, Mercer KL, Adams CR et al (2014) Application of genetic diversity-ecosystem function research to ecological restoration. J Appl Ecol 51:339–348. https://doi.org/10.1111/1365-2664.12202

Ministry of Environment and Forestry (2018) Managing peatlands to cope with climate change: Indonesia experience. The Ministry of Environment and Forestry Republic of Indonesia, Jakarta

Ogden R, McGough HN, Cowan RS et al (2008) SNP-based method for the genetic identification of ramin *Gonystylus* spp. timber and products: applied research meeting CITES enforcement needs. Endanger Species Res 9(3):255–261. https://doi.org/10.3354/esr00141

Page S, Hosciło A, Wösten H et al (2009) Restoration ecology of lowland tropical peatlands in Southeast Asia: current knowledge and future research directions. Ecosystems 12(6):888–905. https://doi.org/10.1007/s10021-008-9216-2

Reynolds LK, McGlathery KJ, Waycott M (2012) Genetic diversity enhances restoration success by augmenting ecosystem services. PLoS One 7(6):e38397. https://doi.org/10.1371/journal.pone.0038397

Ritung S, Wahyunto, Nugroho K et al (2011) Peta lahan gambut Indonesia skala 1: 250.000. Indonesia Center for Agricultural Land Resources Research and Development, Ministry of Agriculture, Bogor

Ritzema H, Limin S, Kusin K et al (2014) Canal blocking strategies for hydrological restoration of degraded tropical peatlands in Central Kalimantan, Indonesia. Catena 114:11–20. https://doi.org/10.1016/j.catena.2013.10.009

Sumarhani, Tata HL (2018) Socio-economic characteristics of community living in peatlands and their perceptions on peatland management: a case study of Jabiren, Central Kalimantan

Province. In: Proceedings of IUFRO–INAFOR Joint International Conference 2017, Yogyakarta, 24–27 July 2017, pp 575–582

Tata HL, Susmianto A (2016) Prospek paludikultur ekosistem gambut Indonesia. Forda Press, Bogor

Tata HL, van Noordwijk M, Jasnari et al (2016) Domestication of *Dyera polyphylla* (Miq.) Steenis in peatland agroforestry systems in Jambi, Indonesia. Agrofor Syst 90:617–630. https://doi.org/10.1007/s10457-015-9837-3

Tata HL, Muchugi A, Kariba R et al (2018) Genetic diversity of *Dyera polyphylla* (Miq.) Steenis populations used in tropical peatland restoration in Indonesia. Mire Peat 21:01. https://doi.org/10.19189/MaP.2017.OMB.269

Thomas E, Jalonen R, Gallo L et al (2014a) Introduction. In: Bozzano M, Jalonen R, Thomas E et al (eds) Genetic considerations in ecosystem restoration using native tree species. The state of the world's forest genetic resources: thematic study. Food and Agriculture Organization of the United Nations, Bioversity International, Rome, pp 3–12

Thomas E, Jalonen R, Loo J et al (2014b) Genetic considerations in ecosystem restoration using native tree species. For Ecol Manag 333:66–75. https://doi.org/10.1016/j.foreco.2014.07.015

Thomas E, Jalonen R, Loo J et al (2015) Avoiding failure in forest restoration: the importance of genetically diverse and site-matched germplasm. Unasylva 245(66):29–36

Wahyudiningsih TS, Naiem M, Indrioko S et al (2014) Allozyme variation of the endemic and vulnerable *Dyera lowii* Hook.f. in Central Kalimantan: implications for genetic resources conservation. Indones J Biotechnol 19(1):79–90. https://doi.org/10.22146/ijbiotech.8637

Widyatmoko AYPBC, Aprianto (2013) Keragaman genetik *Gonystylus bancanus* (miq.) Kurz: berdasarkan penanda RAPD (Random amplified polymorphic DNA). Jurnal Pemuliaan Tanaman Hutan 7(1):53–71. https://doi.org/10.20886/jpth.2013.7.1.53-71

Wilson D, Blain D, Couwenberg J et al (2016) Greenhouse gas emission factors associated with rewetting of organic soils. Mire Peat 17:4. https://doi.org/10.19189/MaP.2016.OMB.222

Chapter 11
Interests Arrangement in the Implementation of Indonesian Sustainable Palm Oil Certification: Case Study of Sari Makmur Palm Oil Smallholders in Riau Province

Bondan Widyatmoko

Abstract The implementation of the Indonesian Sustainable Palm Oil (ISPO) certification remains unsettled nearly a decade since its introduction in 2011. Only a small number of palm oil companies and smallholders have been ISPO certified. Palm oil smallholders became the target of ISPO certification because of their significant contribution to the Indonesian palm oil industry. This study aims to analyze the slow ISPO implementation process at the smallholder level by using a polycentric governance framework, which views the ISPO's overarching rules as changing the relationship between companies and smallholders toward sustainable management. This study examines a smallholder's ISPO implementation in Sari Makmur. This smallholder became a pilot project site for the smallholder ISPO certification in 2011; however, it has still not been certified. This study reveals that the recurring problem of land swaps and the expectations of being certified have complicated the fulfillment process of legal documents for ISPO certification. Nevertheless, smallholders and companies are in the process of delineating their interests toward defining their physical boundaries for sustainable palm oil production.

Keywords ISPO · Smallholder · Polycentric · Land swap · Arrangement

11.1 Introduction

In 2011, the Indonesian government launched a national sustainability certification policy for palm oil commodities, which was called the Indonesian Sustainable Palm Oil (ISPO) policy. Palm oil companies operating in Indonesia must comply with ISPO sustainability principles. Four years into its implementation, this first version

B. Widyatmoko (✉)
National Research and Innovation Agency, Jakarta, Indonesia
e-mail: bond001@brin.go.id

© The Author(s) 2023
K. Mizuno et al. (eds.), *Vulnerability and Transformation of Indonesian Peatlands*,
Global Environmental Studies, https://doi.org/10.1007/978-981-99-0906-3_11

of ISPO was changed through the Ministry of Agriculture Regulation No 11/Permentan/OT.140/3/2015 with regard to the certification system. By 2016, only around 30% of the 1688 operating companies had met the ISPO requirements. These companies had planted almost 1.7 million ha of oil palms and produced 8.9 million tonnes of certified palm oil. These comprised 14% of the total plantation area in Indonesia and 28% of the total palm oil produced in Indonesia (Komisi ISPO 2017). Four years on, this figure has not changed considerably; the ISPO Commission has certified 607 companies or 35% of total operating companies, ten independent cooperatives, and four plasma cooperatives. The low participation in ISPO certification has been attributed to the ministry's weak enforcement mechanisms, which have been unable to solve the many problems around palm oil production. In response, President Joko Widodo released Presidential Regulation No. 44/2020, which seeks to strengthen the ISPO certification system.

ISPO certification is obligatory for palm oil companies, but it is still voluntary for palm oil smallholders[1] until 2025. Through the latest presidential regulation, palm oil smallholders will be obligated to obtain ISPO certification 5 years henceforth. The regulation urges palm oil smallholders to be ISPO certified in view of their significant contribution to palm oil production. Indonesian palm oil smallholders held 4.65 million ha or around 40% of the total palm oil plantation land in Indonesia in 2015. This figure had increased by almost 50% compared to the previous decade (BPS 2016).

There were initially four pilot projects for smallholder certification announced by the government in 2011, all of which are located in Riau Province. Three of the communities involved were supervised by the Astra Plantation Group (PT KTU and PT. SLS). The other pilot project was under the supervision of Asian Agri, with the cooperation of the World Wildlife Fund (WWF). Of these four projects, only one community that was supervised by Asian Agri, and WWF was ISPO certified, indicating the tedious ISPO certification process for smallholders. This article focuses on a smallholder in Sari Makmur supervised by PT Astra that has not received ISPO certification. This pilot project community is selected because the community has a problem of land conflict, which ISPO as a government regulation as well as a sustainability certification system could serve to mitigate and solve. The

---

[1] The definition of a smallholder in this article follows that of the Ministry of Agriculture Decree 98/2013, categorized by landownership of less than 25 ha. Using this definition provides a better perspective for analyzing the general typology of Indonesian smallholders, which consist of plasma and independent smallholders, in terms of their fluid relationship with companies. This fluid relationship is caused by better smallholder terms not only in rising land ownership and national production but also in international entities, such as the European Union, giving attention to smallholders' and local governments' important roles in supporting sustainable palm oil production. These better terms enhance smallholders' capacities for tapping information and accessing commodity pricing, factory production, and government programs. It allows plasma smallholders to negotiate and renegotiate their partnerships with companies and independent smallholders to navigate their interests alongside profitable trading chains and cooperation with plasma smallholders. Observing this process means understanding smallholder capital accumulation without tightly binding it to the existing smallholder typology.

pilot project that received ISPO certification did not have the land legality problem. Therefore, the process to get ISPO certification was mainly focused on sustainability management in the palm oil plantation.

This study seeks to analyze the reasons behind the slow ISPO implementation process for smallholders. In so doing, this study raises two questions. First, what factors account for the slow process of ISPO certification at the level of palm oil smallholders? Second, how have the smallholders responded and overcome this problem?

This study uses the polycentric governance system to analyze smallholders' self-governance process to obtain the certification. This chapter analyzes the capacity of the smallholder community to influence the outcome of ISPO implementation through its existing economic and capital accumulation, political relations, and cultural infrastructure. This condition has allowed it to make arrangements for negotiating or compromising on its interests with companies and/or local government to solve its problems. The community's success in cooperation and ability to provide solutions are the notable characteristics of smallholder polycentric governance (Carlisle and Gruby 2019).

This study is organized into six sections. Section 11.2 reviews the literature to show the significance of this study. Section 11.3 discusses the methodology, which consists of the research site, scope, and data collection. Section 11.4 discusses the conceptual framework and hypothesis of this study. Section 11.5 discusses findings and a discussion on how smallholders, the local government, and the company try to overcome the problems to fulfill ISPO certification requirements, including making institutional arrangements through communication and negotiations. Section 11.6 concludes the study.

11.2   Literature Review

In the development of commodity certification, the certification is not only merely about quality control; it has become a social and environmental institution. Commodity certification has been endorsed amid the doubt over governments' effective regulatory mechanisms to mitigate severe environmental degradation and the prevalence of social injustices. Limited budgets and a lack of human resource capacities are commonly blamed for government bureaucracy. Commodity certification is generally implemented by private entities to build governance systems that can avoid such asymmetric information problems, take immediate measures, and adjust accordingly. Institutionally speaking, commodity certification sets operational compliance standards to meet principles, such as environmental protection or labor rights, and it provides a chamber for multistakeholders to participate in decision making, monitoring, and reporting processes. The certification does not have the coercive power that government regulations do. However, it does depend on market mechanisms—in this case, the public consumer—to respond to institutional misconduct. A public consumer can place pressure on commodities in the form of protests, "blaming and shaming" strategy campaigns, and boycotts. Multiple stakeholders within the certification governance can then take measures accordingly (Cashore et al. 2008; Vandergeest 2007; Ebeling and Yasué 2008; Bartley and Smith 2010).

The performance of certification mechanisms for social and environmental institutions has been put into the spotlight of contention. Some literature argues that commodity certification could bring welfare to a targeted community through increasing commodity prices and a certain degree of financial freedom for household income allocation (Bacon 2005; Elbers et al. 2015; Hidayat et al. 2015; Cameron 2017). Although this goal might not be easy to reach, other researchers have argued that community organizational factors, such as preexisting social capital (Bray et al. 2002; Pérez-Ramírez et al. 2012), further factor production incentives and managerial support (Suradisastra 2006; Ruben and Zuniga 2011; Beuchelt and Zeller 2012), and prior experience in commodity certification (Bacon 2005) are necessary factors to consider. These factors are needed to maintain organizational and institutional consolidation, community negotiation capacities, and community resistance against external shocks.

However, other studies express that commodity price incentives given to farmers have the potential to co-opt a community's preexisting social values through crowding out processes (Frey and Obelhozer-Gee 1997), incorporation and exclusion (Somerville 2000; Li 2007; McCarthy 2011; Dove 2012; Peluso 2016). In a traditional community in which organization and cooperation are built based on necessity and reciprocity principles, the presence of price incentives in a commodity certification organization could lead to new values for measuring relationships in the community and crowding out prior social values. Guthman (2004) depicts these shifting values in cooperation as rent-seeking activities by communities, which can further drive away initial aims or values of commodity certification, such as sustainability and fairness. Gamson (2005) further describes a situation in which the missing new advantage to community acceptance for certification is a form of co-optation.

Similarly, researches on ISPO are conducted by comparing aspects of sustainability governance of ISPO with the Roundtable on Sustainable Palm Oil (RSPO) and typically found that the ISPO certification per se did not promote sustainability aspects (Brandi et al. 2013; Jiwan 2012; Suharto et al. 2015; Potter 2016; Rietberg 2016; Varkkey 2016; Erman 2017; Hutabarat 2017; Hidayat et al. 2018). In addition, other research has shown that certification also cannot guarantee market access and higher prices (premium prices) as promised by RSPO certification (Hidayat et al. 2015; Cramb and McCarthy 2016) as well as struggling with accountability problem (Jiwan 2012). This condition would make the ISPO certification even harder to obtain, as the global market has been reluctant to approve its operation (Hidayat et al. 2018). In contrast to this research, several studies have shown that certification will guarantee market access, better prices, incentives for oil palm farmers (Frey and Obelhozer-Gee 1997; Cameron 2017), redistributive benefits through free prior inform consent mechanism (McCarthy et al. 2012) and a form of government paternalism policy and inclusion (Silver 1994; Werle 2001; Wolf 2006; Thomas and Buckmaster 2010).

Nonetheless, there are drawbacks to these studies. The research on palm oil certifications mentioned above has always been carried out on certified oil palm smallholders. These smallholders did not face land problems or conflicts, so the findings may not be representative of the obstacles faced by palm oil smallholders in Indonesia, who were mostly beset by land problems or conflicts in the process of obtaining an ISPO certification. Therefore, this study fills the gap by showing the process of securing ISPO certification by palm oil smallholders who in fact still face problems, such as with land tenure and social conflict. This study shows to what extent ISPO certification can induce initiatives and local arrangements to overcome obstacles in the implementation of sustainable palm oil certification.

11.3 Research Site and Methodology

11.3.1 Research Site

The village of Sari Makmur, previously called Satuan Pemukiman 6 or SP6, is located in the Pangkalan Lesung Sub-District of Pelalawan Riau, which had been formed together with the opening of the palm oil plantation of Sari Lembah Subur Company (PT. SLS) in the late 1980s. This village was named a pilot project for ISPO certification for smallholders.

The total village settlement covers approximately 300 ha and 1000 ha of the total plasma plantation area. The village became a plasma area for PT. SLS. The community of this village comprises migrants from other islands, especially Java Island, from the three waves of transmigration for palm oil development through the nucleus-plasma scheme (Trans-PIR) in 1989, 1990, and 1991 (see Table 11.1). The demographics have since changed due to the Trans-PIR families' descendants and people's movements (see Table 11.2).

As smallholders in a nucleus-plasma system, each household received around 2.5 ha of land—2 ha to be used for palm oil cultivation (plasma) and 0.5 ha for settling and growing a living source of crops. The company prepared the 2-ha plantations for this plasma, including the seed and fertilizer. The smallholders had to pay the company back for this setup after the fruits were ready for harvest. It took 8 years for them to pay off the Rp9 million debt that started in 1994—4 years sooner than projected.

Newcomers were those who arrived in this village after its independence from the Ministry of Transmigration's administrative responsibility in 1997. These newcomers, who worked as wage laborers, were motivated by promising job offers and family reunification. Once the laborers became financially independent, they bought a piece of land for their own palm oil plantation. The smallholders of Sari Makmur also benefited from the presence of seasonal laborers who worked on their plantations. This type of labor usually came from other districts in Riau Province, such as Bengkalis or Meranti. They only worked for several months and did not stay on afterwards.

Table 11.1 Initial transmigrant flows and settlements in Sari Makmur (Source of data: Village administration documents and interview on September 27, 2017)

No.	Year of arrival	Area of origin	Household number
1	1989	Central Java	50
2	1990	Yogyakarta	50
3	1990	Central Java	50
4	1990	Yogyakarta	40
5	1990	DKI	50
6	1990	East Java and Madura	50
7	1990	Local allocation transmigration settlement (Alokasi Penempatan Penduduk Daerah Transmigrasi, APPDT)1	20
8	1990	APPDT 2 (from Genduang Sub-District)	36
9	1990	DKI Jakarta	10
10	1991	Independent transmigration (Swakarsa)	104
11	1991	APPDT 3	40
Total			500

Table 11.2 Sari Makmur household status and village area distribution (Source of data: Village administration documents and interviews, 2017)

Neighborhood and community association	Village ward organization	Household status			
		Trans-PIRs	Trans-PIR household descendants	Newcomers	Moved out
RT 01 RW 01	Agung Mulyo	20	15	–	–
RT 02 RW 01		20	23	10	–
RT 01 RW 02		6	3	9	–
RT 02 RW 02		17	5	23	–
RT 03 RW 02		10	4	18	–
RT 01 RW 03	Karang Sari	7	8	22	–
RT 02 RW 03		20	20	18	–
RT 01 RW 04		9	2	15	–
RT 02 RW 04		21	4	29	1
RT 01 RW 05	Rejosari	15	4	17	–
RT 02 RW 05		21	6	18	2
RT 01 RW 06		22	4	14	–
RT 02 RW 06		22	4	16	–
Total		210	102	209	3

11.3.2 Methodology

The research scope is limited to the period between the implementation of ISPO for the first time in 2011 under President Susilo Bambang Yudhoyono until the final research conducted in 2017 under the President Joko Widodo.

This study uses quantitative and qualitative methodology with data collected from primary and secondary sources. The primary data was collected through a questionnaire and interviews with 50 households in Sari Makmur Village, Pelalawan District, Riau Province, from August until November 2017. These 50 households were chosen through a random sampling method. Once those chosen agreed to be respondents, they were provided with a questionnaire that consisted mostly of open-ended questions on the household's historical formation, migration processes, land ownership history, palm oil production, and other information concerning household members' relationships with the company and local government. The primary data on capital accumulation was extracted from a national land certificate owned by a Sari Makmur household, which provides financial banking transaction records. These records, usually called *catatan roya*, hold information on Sari Makmur households' past debt transactions registered by their notary and acknowledged by the National Land Agency (BPN). From the 500 available certificates, 377 were successfully listed. Approximately 216 of the collected certificates were used by the owners as collateral for bank loans. Primary data was also collected through interviews with the local government of Pelalawan District, the ISPO commission of the Ministry of Agricultural, and the UNDP representative in Jakarta to cross-check information regarding ISPO implementation at village level. Secondary data was collected from documents, regulations, books, journal articles, and government reports related to the implementation of ISPO.

There are two steps to data analysis. First, the interview record is transcribed. Second, the answers, which have been grouped, are analyzed through analytic induction. Eventually, all the data gathered from the written sources and interview are interpreted and analyzed to produce a claim within the framework of the research question.

11.4 Governing Palm Oil Sustainability Through Polycentric Governance and ISPO Certification

This study utilizes the polycentric governance perspective to observe the ISPO implementation, an approach that is particularly beneficial for analyzing the institutional complexity of the natural resource system's situation, improving adaptive capacity, and mitigating risk on overlapping governance actors and institutions (Marshall 2015; Carlisle and Gruby 2019). But some drawbacks, such as complex accountability, leakage, redundancy, high transaction cost, and free-rider problems, can be found in polycentric governance implementation (Marshall 2015; Morrison

2017). Nonetheless, the perspective this approach takes provides a substantial scenario for analyzing institutional arrangements in addressing social problems, especially resource management, as well as relationships and interactions among the actors involved (Thiel et al. 2019). In order to reach such an outcome, this approach requires the exploration of three essential elements of polycentric governance.

> Polycentric governance connotes many centers of decision making that are formally independent of each other. Whether they actually function independently or instead constitute interdependent system of relations is an empirical question of particular case. To the extent that they take each other into account in competitive relationships, enter into contractual and cooperative undertakings or have to recourse to central mechanisms to resolve conflicts. (Ostrom et al. 1961).

Based on the statement above, the first element of polycentric governance is the independence of each actor/unit involved in resource management. Marshall (2015) provides an interpretation of independence, which refers to de facto autonomy from one to another (Marshall 2015; Schröder 2018). This means that each actor/unit has a certain degree of freedom in identifying attributes, developing rules and sanctions, and carrying out a decision-making process independently. According to Marshall (2015), this interpretation is more functional than taking the perspective of de jure autonomy since possessing de facto autonomy would still give the opportunity and ability to arrive at a working arrangement with other overlapping actors/units in resource management, even in a situation in which de jure autonomy is absent (Marshall 2015). In addition to this interpretation, if the actor/unit's independence is characterized mainly by de jure autonomy, they tend to be exclusive and concentrate on maintaining their existence in the face of the overarching rules prescribed, especially by the government. This does not mean that recourse to the process of rules is abandoned. However, in a situation in which an actor/unit lacks legal knowledge and access, strengthening de facto autonomy is urgently required.

The second element is the interdependent relationship between these actors; despite their individual independent nature, each actor is linked to a certain type of relationship arrangement in managing the resources. This relationship is formed because of overlapping jurisdiction among actors/units in resource management. The overlapping jurisdiction is the engine for actions, reactions, and interactions among actors/units involved in resource management (Stephan et al. 2019). There are three sources of overlapping jurisdictions: (1) interdependence between the issues being governed; (2) the interconnectedness of the physical jurisdiction; and (3) the function of governance, such as monitoring (Stephan et al. 2019). Schröder (2018) suggests that for the overlapping jurisdiction to become more functional, the relationship should focus on good aspects or problems (points 1 and 2) that connect actors/units involved in resource management.

These overlapping jurisdictions attributed to actors/units would describe the third element of the types of relationships in polycentric governance. There are three recognizable relationships, namely, cooperation and collaboration, conflict, and competition (Ostrom 1990). Koontz (2019) provides some of the important factors that could either incentivize or hinder cooperation and collaboration. The first factor is authority, which is marked by the lack of vertical hierarchies and power imbalances. This situation could lead to a dual scenario of conflicting engagement among actors/units but also cooperation and collaboration as a response to power imbalances. The latter situation might require the actors/units to share values, beliefs, and goals toward a common understanding and interest. To achieve this, the second factor, which is responsible information, needs to be circulated. This could lead not only to the process of learning but also trust among actors/units. The third factor is resources, which capacitates collaboration among actors/units. Koontz (2019) further notes that the lack of resources can incentivize actors/units to join forces for collaboration. It should also be considered that the source of funding could affect collaborative partnerships, especially when funding is limited.

Power imbalances, incomplete information, and insufficient resources are some factors that hinder collaboration and tend to lead to conflict among actors/units. However, in a polycentric governance, actor/unit adaptation and learning capacity in a conflict situation through conflict-resolution cycle could also catalyze further collaboration (Hekkila 2019). This learning capacity and adaptation in a conflict situation could also improve actor/unit efficacy toward exploring opportunities and channeling rules for better resource appropriation, provision, and production of public goods (Garrick and Villamayor-Tomás 2017).

In the context of ISPO implementation, the government has emphasized sustainability as an overarching condition of certification for both companies and smallholders. The ISPO certification thus intends to transform production relationships between companies and smallholders and their perspectives and initiatives in favor of the environment. Several regulations inserted in the ISPO sustainability principles ostensibly compel company and smallholder operations to comply with sustainable palm oil production (see Table 11.3).

The ISPO upholds the legal aspect as the first principle in building sustainable palm oil production before plantation management, conservation, environmental surveillance, labor, and social responsibility and business improvement. The legal aspect is the gateway to arrange and rearrange licenses and permits for companies and smallholders in palm oil production. Legality almost always ensures that other sustainability principles will be easily fulfilled.

The ISPO also has a direct mechanism for sanctions, including a warning notification, fine, temporary suspension of operations, putting the certification on hold, and revoking it altogether. To some extent, the ISPO has indirectly forced companies to strengthen partnerships with smallholders—for example, companies source a minimum of 20% palm oil supply from smallholders. However, ISPO certification has embedded inclusive benefits, which give government and financial institutions the assurance of legality, sustainability, and profitability. This assurance is positively related to access for aid and assistance to improve palm oil production.

Within this scheme, the ISPO has provisions for both the company and the smallholder competing to obtain certification. However, it should be noted that palm oil development does not happen in a vacuum; it is often backdropped by stories of conflict and environmental degradation. Such stories are characterized by power imbalances between the company and community, as well as imperfect

Table 11.3 Principles of ISPO certification for palm oil companies and smallholders (Menteri Pertanian Republik Indonesia 2015; Presiden Republik Indonesia 2020)

No.	Description	ISPO certification		
1	Object of certification	*Company*	*Smallholder*	
		Plantation	*Plasma*	*Independent (Swadaya)*
2	Scheme	Mandatory	Mandatory by 2025	Mandatory by 2025
3	Object of audit	• Company	• Manager • Cooperative/ farmer group • Farmer (owner or worker) and their farm	• Cooperative • Farmer group • Farmer (owner or worker) and their farm
4	Criteria	• Plantation legality • Plantation management • Primary forest and peatland conservation • Environment management and surveillance • Labor responsibility • Social responsibility and economic empowerment of the community • Improvement of sustainable business	• Legality of smallholder (STDB) • Management of plasma plantation • Environment management and surveillance (SPPL) • Working safely and responsibly • Social responsibility and community empowerment • Improvement of sustainable business	• Legality of smallholder (STDB) • Farmer organization and management of business • Environment management and surveillance (SPPL) • Improvement of sustainable business
5	Requirements	• Plantation license permit (IUP/IUP-P/ IUP-B/SPUP) • Land use right (*Hak Guna Usaha*, HGU) • Environmental permit • Class of plantation determination (I/II/III) from the Head of District/Major/ Governor/Director General of the plantation	• Document of plantation formation • Copy of ISPO Certificate from Nucleus Plantation (*Kebun Inti*) • List of the group members' names • Right of the lands: Land ownership (*Surat Hak Milik*, SHM)	• Document of cooperative/farmer group formation • List of the members' names • Land ownership such as *Surat Hak Milik,* letter C/ *Girik, akta jual beli* or other legitimate documents

information. These in turn create various descriptions of community co-optation, which lead to patron–client relationships between the company, *tauke* or middleman, and smallholder. Moreover, imperfect information influences the community's access to capital, production, and legal knowledge. This has the effect of putting the community at a disadvantage, as it is unable to improve productivity and its legal position in a conflict situation.

However, polycentric governance gives another perspective, one in which a power imbalance and imperfect information can induce cooperation among the actors, especially in palm oil production. Van der Enden (2013) has shown that palm oil development through government programs has not only empowered communities in terms of improving local economies but has also opened pathways for sustainable policy arrangements. Her research acknowledged that intensified palm oil development methods introduced by the government had been adopted by one of the communities in Siak Riau (van der Enden 2013). This community's commitment to sustainability has led to other conservation initiatives involving both national and international nongovernment actors.

11.5   Findings and Discussion

Based on the theoretical framework previously discussed, the first element of polycentric governance in this case comprises the palm oil smallholders in Sari Makmur village incorporated in the Mulia Cooperative (Mulia). The existence of Mulia as an economic entity maintains the resource appropriation and capital accumulation processes. The current Mulia cooperative not only binds the community's palm oil economic activities but also empowers the community to realize the value of the legal standing of their resources, especially to access financial assistance. This condition enables the smallholders in Sari Makmur to be independent actors to make decisions in the production and cultivation process, distribution system, partnership with the company, and participation in local government programs.

Another independent actor is PT. SLS as a palm oil company. This company operated as a nucleus in the nucleus-plasma program in the 1980s. This status changed after the program completed in the 2000s. The company no longer has an obligation and responsibility to its plasma. The plasma smallholders in Sari Makmur do not have any obligation to the company, either. Therefore, both PT. SLS and smallholders are independent and free to determine their relationships.

Both these independent actors have built a partnership based on interdependent interests in managing the palm oil plantation. This interdependent relationship is the second element of polycentric governance. On one side, PT. SLS cannot rely on its core plantation to supply fresh fruit bunches (FFB) of palm oil. The company needs an FFB supply from the palm oil smallholders to fulfill its daily target of processing production. On the other side, the smallholders have the same interest. They need the company to buy their fruits for a good price, even though they have the freedom to sell their produce elsewhere. The smallholders require factory production that is readily available, stable pricing, and information on good production management, all of which are usually provided by the company. This interdependence of both actors is an incentive for cooperation between PT. SLS and palm oil smallholders in

Sari Makmur. Nonetheless, both actors also have a conflicting relationship regarding land status legality due to past land swap processes.

In relation to palm oil sustainable management, the Indonesian government has implemented ISPO certification. This certification is mandatory for companies and will be an obligation for smallholders in 2025. SLS is a company that holds ISPO certification. Meanwhile, a smallholder in Sari Makmur village is a pilot project for ISPO certification due to its successful plantation management under Mulia. For this achievement and its past participation in a government program on nucleus-plasma palm oil development, this smallholder is considered for having no peculiarities on land legalities. However, this smallholder has not been awarded ISPO certification. Why has it not received the certification 10 years after it became the pilot project? I found that the problem of land swaps caused a legality issue with its land status. This problem is hindering the smallholders in obtaining a smallholder certificate for plantation registration (*Surat Tanda Daftar-Berkebun*, STDB) and statement letter for environmental management (*Surat Pernyataan Pengelolaan Lingkungan*, SPPL) documents as a requirement for certifying ISPO. Another obstacle in the process of smallholder certification are the unclear procedures to obtain STDB and SPPL documents under the local government authority.

How, then, do the palm oil smallholders tackle these obstacles? Being certified by ISPO is an important issue for smallholders in Sari Makmur because ISPO certification is a precondition for them to be eligible to receive the government replantation fund, which important for their palm oil replantation. The implementation of ISPO in the community also promises to strengthen this social capital since it would help them progress to the process of replanting. This condition of a smallholder's old plantation is problem for the company due to the potential disruption of FFB supply. If the replantation program is not successful, it will also influence the company. In addition, holding ISPO certification gives PT. SLS a reason to maintain a good relationship with the community. Therefore, if conflict occurs, it is obligatory for the company to solve it. Otherwise, the ISPO certification would be forfeited by the government.

Due to these reasons, based on the third element of polycentric governance, both the company and community have to make arrangements to solve those problems to fulfill the sustainable requirement on ISPO certification. PT. SLS and smallholders in Sari Makmur negotiated initial arrangements to complete the certification process instead of awaiting sanctions from the government for their failure to fulfill ISPO requirements. Both parties are working together to complete the certification procedure, which also means finding solutions to land swap issues and obtaining STDB and SPPL documents for smallholders. The current dynamics, especially in completing the legal process, see both parties putting their resources toward a definition of their physical boundaries in pursuit of sustainable palm oil production.

11.5.1 Self-Governance Through Mulia and Capital Accumulation

The village of Sari Makmur was one of four pilot project areas in Riau for ISPO implementation. Its local government had officially announced this status in January 2015. These pilot project areas are expected to show the way for other smallholders to obtain ISPO certificates. Mulia here is the main agency responsible for ISPO implementation. It deals with administrative documents for registration and maintains sustainable practices in plantations following certification. The local government has acknowledged the cooperatives in these four villages as the best financial performers. Mulia works with PT. SLS, which has been ISPO certified.

Mulia was founded together with the administrative formation of Sari Makmur in 1997. Mulia manages 34 farmer groups, which originally organized 500 households that owned a plasma plantation. However, the actual number of households today is less than 500 because of ongoing land transactions. From PT. SLS land certificate data, it was found that 38 of these households possess more than one plasma land certificate. Therefore, only around 462 members are currently active.

Mulia functions as a bridge for constant communication regarding issues related to plantation management faced by the farmers group. The farmers' group leader is obligated to deliver the meeting results to the group members. Everyone in the group has rotational responsibilities for harvesting, loading, and maintaining security in the smallholder oil palm plot (*ronda*). All Mulia members are also obligated to attend the annual meeting to hear the report from their leader and administration on the cooperative's performance.

Mulia also helps the community to get easier access to affordable daily goods and plantation necessities by providing a grocery store because the village is in a remote area. In addition, it provides a transport business for its members to deliver palm oil fruits to processing companies.

Mulia also provides its members with recommendations when they want to access the bank's loan facilities. The bank needs Mulia to double-check the documents, incomes, and the smallholder's ability to repay. This is why Mulia is vital for palm oil smallholders in Sari Makmur, especially for capital accumulation and land expansion.

Sari Makmur community members generally work either as owners or laborers of their palm oil smallholdings. Landholders regularly nurture palm oil productivity. They check the availability of fertilizer stock, field distribution and labor, and the palm oil's harvesting schedules and current prices. Such information is verified through phone calls or visits to the cooperative. Harvesting, fertilizing, and pruning are usually handled by available wage laborers in the village. This kind of work requires strength and knowledge about palm oil trees, especially when a tree's height is at or above 10 m, and the tree is more than 20 years old. The owner usually checks the work of their laborers in the fields but prefers to do other regular field jobs themselves, such as *babat* (slashing palm oil weeds) and *brondol* (collecting loose fruit from the trees).

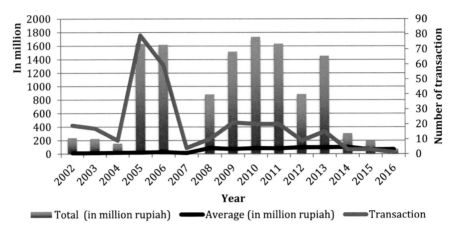

Fig. 11.1 Financial banking records from Sari Makmur Community Land Certificates. Source of data on Transmigrant National Land Certificate were collected and compiled by the author during fieldwork in Sari Makmur Village

Plantation jobs are both physically and mentally intense for laborers who do not have landholdings. They have to finish their jobs on time and with satisfying results. For example, a landholder will personally contact laborers several days before the scheduled harvest to ascertain their availability. On harvest day, laborers start working in the early morning and have to finish harvesting before it's time to transport the harvest. Usually, harvesting is completed in 2–3 h for 2 ha of land. Laborers must have good eyesight and cutting skills to pick and slice the palm oil fruits that are ready for harvest. Mistakes such as choosing over- or underripe fruits and wrongly slicing them will cost the owner time and money and reduce palm oil tree productivity. Quality laborers are usually paid around Rp150,000 per 2 ha, excluding lunch and a pack of cigarettes. Subpar performances merit several warnings from owners and, ultimately, termination (sometimes without notice) in case of consistently bad results.

The working sector in this village (besides the palm oil industry) is limited, as other sources of work have not become an option for the general community. This has made the community rely solely on palm oil as its source of income. Unfortunately, with aging palm oil trees, the community faces challenges regarding its future income; however, it manages this risk through land investment. Thus, when replanting time approaches, the community already has an alternative source of income. This strategy is seemingly easier for people to take since financial services from banks are readily accessible. After community members, especially Trans-PIR households, receive land certificates from the bank, they continue to access the institution's financial support for many necessities (see Fig. 11.1), which mostly involves buying land (see Fig. 11.2). This condition is supported by the availability of land in the community and in nearby areas.

Figure 11.1 shows the community's interactions with financial institutions from 2002 to 2015. As it demonstrates, the highest number of banking transactions—84—

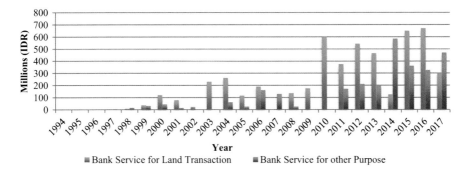

Fig. 11.2 Sari Makmur community bank loans by borrowing motives (Source of data: Survey in Sari Makmur, 2017)

took place in 2005, particularly with Bank Negara Indonesia. From these transactions, the community's average proposed loan was Rp19 million. These proposed loans quadrupled, reaching an average of around Rp80 million some 4–5 years later.

With debt payoff usually taking around 3–4 years, the number of banking transactions should have risen. However, there were some institutional barriers that checked the desire for capital accumulation. Since mid 2000, the community had not easily been able to interact directly with the bank to propose loans. The community needed to show the bank written permission from the farmer groups' leaders, as well as cooperative seals. This written permission became the guarantee, which stated that a member was financially capable of repaying the debt. It was used to prevent loan defaults. Additionally, since 2013, the bank has reduced the amount of financing for the plantation. Despite this, the village's financial institution (*Badan Usaha Milik Desa*, BUMDes) has been performing very well in Sari Makmur. BUMDes can provide loans below Rp20 million for 2 years. Unlike the bank, BUMDes's fast service and easy conditions for loan approval enable the community to access financing from it.

Since the *roya* notes did not mention specific reasons why the community accessed bank services, in-depth household interviews were carried out from August to November 2017 to clarify the community's motives. The interviewers asked questions mostly regarding families' formation histories, banking interactions, purchased land, ownership, and palm oil production. Fifty households were randomly selected in Sari Makmur to participate in these interviews.

Figure 11.2 indicates that Sari Makmur families had tended to borrow from the bank for land-related transactions. Before 2000, less than Rp100 million accumulatively was used to buy land. However, it appears that in the last decade, the accumulation tended to grow to above Rp400 million, reaching its highest at Rp670 million in 2016. This acceleration of capital accumulation was not solely because of the many land transactions that had been conducted by families. It was also because of the high price of land around the village. Approximately 2 ha of land comprising hills and with less infrastructure could cost Rp250–300 million. Many communities said that even at those prices, people still lined up to buy land. In fact,

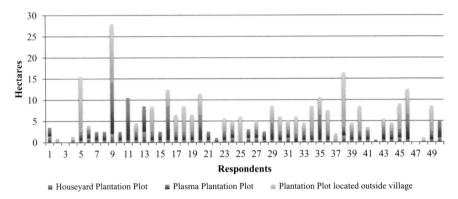

Fig. 11.3 Sari Makmur community land ownership area by location (Source of data: Survey in Sari Makmur, 2017)

the families that were surveyed reported that their highest land transactions had occurred around 2004. In the last 5 years, Sari Makmur families have also tended to use bank loans for other purposes, such as marriage ceremonies, houses, cars, and financing their children's higher education in Java. The highest accumulation for loans unrelated to land transactions reached Rp585 million in 2014, which was 14 times greater than that of 1997.

The community divides its plantation land into three locations: house yards, plasma, and plantations located outside Sari Makmur (see Fig. 11.3). Two decades ago, this distribution was unlikely to be found in the community. Each transmigrant family generally owned 2.5 ha of land. However, it appears that the community's financial interactions with the bank especially helped these households to accumulate land outside the community. Nowadays, each household owns an average of approximately 6 ha of palm oil smallholdings. These consist of a 0.6-ha house yard, 2.5 ha of plasma holdings, and approximately 3 ha of holdings outside the community.

Figure 11.4 shows that plasma holdings comprised the most palm oil production space in 2017. Families claimed that, on average, their plasma holdings could produce a total of 5 tonnes of palm oil per month (Fig. 11.4). This production is still high considering the age of the trees. The main reason for this is the high level of communication among the farmers' groups and the willingness to collaborate within cooperative organizations. Mulia not only functions as an organization of farmers' groups that communicates with the company but is also a venue for regular factory production at an affordable price. In addition, it has become a communications clearinghouse, especially on matters of plantation productivity, labor matters, transportation problems, and commodity prices. Within cooperative organizations, there is moral suasion to maintain the quality and quantity of palm oil production. Therefore, it is typical for members to fertilize their plots three times a year, weed or clean three or four times a year, and tidy up palm oil fronds once a year. With such regular care, farmers need not worry about the *musim trek* (seasonal downtime for

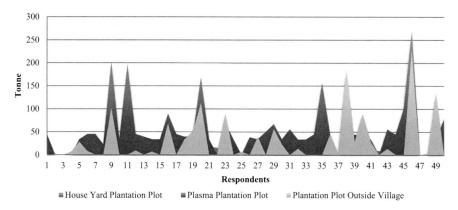

Fig. 11.4 Sari Makmur community palm oil production by location (Source of data: Survey in Sari Makmur, 2017)

palm oil production). Farmers do admit, however, that during this three- to four-month period, their production might decrease, on average, to 20%. In comparison to independent smallholders, the Sari Makmur community has a doubled production figure and a lower percentage of production loss during *musim trek*.

There are several differences between plasma holdings, house yards, and outside-village holdings. First, in terms of production, a house yard and an outside-village holding can produce 1 tonnes and 3 tonnes of palm oil per month, respectively (Fig. 11.4). For outside-village holdings, production is relatively low. With an average of 3 ha of land, an outside-village holding should be able to produce at least 5.5 tonnes per month. Moreover, knowing that palm oil trees usually reach their most productive point at 15 years of age, these figures can be easily exceeded. Although farmers have admitted to implementing similar smallholding management styles as in their plasma, it seems they pay more attention to smallholdings with national land certificates (*Surat Hak Milik,* SHM) rather than land certificates recognized by local leaders (*Surat Keterangan Tanah,* SKT *or Girik*). This priority is indicated in land productivity, in which plasma holdings have the highest production of up to 1.8 tonnes per ha. This is followed by house yard holdings and outside-village holdings, which produce 1.5 tonnes and 0.76 tonnes of palm oil per hectare, respectively.

Second, house yards and outside-village holdings are not under cooperative management. Therefore, they do not directly relate to PT. SLS in terms of price and selling agreements. Farmers are free to choose where they sell their palm oil FFB; price becomes the main factor in this decision. The Sari Makmur community usually sells its produce to a local palm oil collector known as the *tauke*. There are price differences; it is around Rp100–200 cheaper per kilogram to sell to a *tauke* when compared to the PT. SLS price, which is relatively stable at around Rp1500–1700 per kilogram.

Even with this impressive economic achievement, the community is worried about the replanting they should carry out in the next 2–3 years. Without cutting down the old palm oil trees and using better seeds, the community's palm oil productivity will decrease, as well as its income. However, it is also concerned that the replanting process could erode its cohesion as a community and break down its institutional foundations in Mulia. This concern was expressed by some farmers' groups during a monthly meeting. They felt that following the replanting program administered by the company put them at a disadvantage because the cost was too high. They argued that it would be cheaper if they replanted their own way. However, other groups said that individual replanting processes would endanger the environment because of the poison used to kill the trees. Individual replanting also means breaking the partnership contract with the company, which means losing the benefits of a stable and better palm oil price.

11.5.2 Solving the Land Swap Problems Between Smallholders and Companies

The ISPO was introduced to Mulia for the first time by PT. SLS in June 2014. The company invited representatives of plasma smallholders to attend an ISPO socialization. As one of the communities targeted to receive ISPO certification, Mulia and several farmers' organization leaders attended this event. During the event, a representative of the ISPO Commission discussed the definition of ISPO, its benefits, and some processes needed to obtain the certificate. The meeting was followed by a visit to the Mulia office for another round of socialization and interviews with several farmers on the cooperative's current plantation management. After this, Mulia management and PT. SLS started to collect the required documents from the community to apply STDB and SPPL.

Immediately, they faced several difficulties. The problem of document availability as well as the distrust of some community members of document collection became issues during this process. Sometimes, the documents were still in the bank as loan collateral, and other times these documents were lost and therefore could not be provided. In these situations, extra time was needed to get copies or to recover the documents. Another problem was the community's distrust of PT. SLS due to land conflicts that had occurred in the past. This was exacerbated by fraudulent land transactions experienced by some members of Mulia.

To regain the community's trust, PT. SLS became directly involved with Mulia's administration, appointing an official company worker in the Mulia office on a regular basis. Meanwhile, the head of Mulia also assigned village elders to work as cooperative secretaries. These two agencies jointly became a symbol and a strong guarantor for the community to put their trust in the ISPO document collection process. It was important for the community to know that there was an agency they could turn to in case something went wrong—such as the fraud case that the community had been unable to handle on its own. However, it is interesting that not only narratives of ISPO requirements were used during the collection but also replanting matters. By owning STDB and SPPL, not only would the community receive ISPO but it would also be eligible for financial government aid for replanting in its plasma plantation. This aid, which reached Rp25 million per ha, was of greater interest to the community than the ISPO issues. Moreover, the lack of information about ISPO, which only tended to be shared among the co-operative's staff and farmers' group leaders, made ISPO certification a less attractive goal.

Besides the problem of fulfilling the required documents, the process of ISPO certification in the village of Sari Makmur also brought up past land conflicts between the community and PT. SLS. This conflict was related to the plasma land distribution during the early days of Sari Makmur as a transmigrant village. There had been constant elephant attacks on plasma land that encroached on elephant habitat, rendering the land unproductive. The unresolved issues of elephant conservation had compelled the PT. SLS to delay the distribution of several portions of plasma land until conservation matters were resolved, which did not sit well with the community, which demanded that PT. SLS fulfill its obligations sooner. This clash led to protests and harsh demonstrations.

In response, PT. SLS, in cooperation with the local government and related institutions, offered a land exchange to affected families. This included the 47 families who agreed to the offer. These households cultivated block area numbers 31–34 (see the green area in Fig. 11.5). The company and households agreed that this exchange was only temporary and that all of these families would eventually return to their initial distributed plasma areas when the latter could be feasibly cultivated by the community—i.e., within one planting period for the plantation, which is around 25 years.

By the end of the period, however, not all 47 households agreed to return to their initial areas, as 17 households who had cultivated block area number 32 still refused to return to P21 and P22 (see the white areas on Fig. 11.5). Even though these locations had been replanted with new trees, and the households would therefore not have to pay for replanting costs for 25 years, they still had concerns with the areas' poor infrastructure. The households demanded that PT. SLS pay compensation (Rp50 million) to each smallholder to build new infrastructure for their lands. However, PT. SLS has not yet granted this demand, as it does not consider the matter urgent. After all, the other 30 households had already moved and were cultivating block areas P18, P19, and P20 without issue.

In the case of STDB and SPPL applications in the community, ISPO administrative requirements have divided the interests of the smallholders' community and the company. For Sari Makmur, ownership of these documents means access to further capital accumulation through government assistance and funds from financial institution. It also means a guarantee of stable market access, especially for PT. SLS, leading to better commodity prices as well as the legal acknowledgment of their plantations. However, this community first has to solve the problem of land swap with the company in order to get the STDB and SPPL for all smallholders in Sari Makmur.

For PT. SLS, pushing for STDB and SPPL application in the community will pave the way for it to close concessions and plasma plantation border disputes. It is

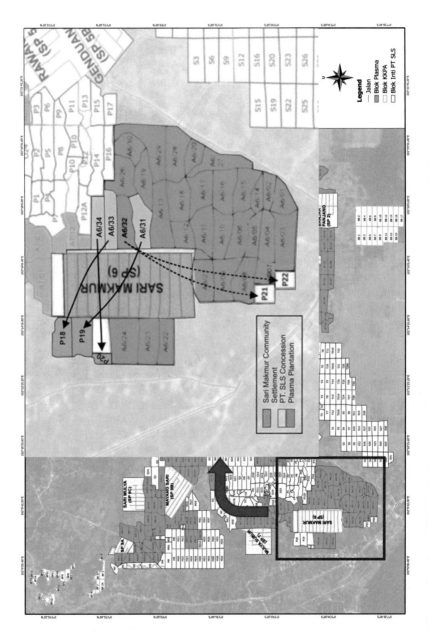

Fig. 11.5 Plasma plantations within PT. SLS land concessions in Sari Makmur (Adapted from Widyatmoko 2019)

important for this company to renew its rights to the palm oil plantation in this area given that the permit will expire soon. An ISPO application by the community would be a great reason for PT. SLS not to grant compensation demands from several smallholders who do not want to return to their original plantations. This is because the ISPO application requires legal acknowledgment of community-cultivated plantations. A community application will be rejected if the ISPO Commission finds that some of the plantations are still located inside the company concession.

Furthermore, the clearance border dispute is also important for PT. SLS, which not only has to renew its HGU but also its ISPO certification by January 15 2019. Unless the company resolves the dispute before this due date, ISPO certification will be revoked. This condition will harm PT. SLS's reputation as a nationalist company determined to use national certification for sustainability. For this reason, it will not be surprising if PT. SLS offers assistance to the community in applying for ISPO certification.

11.5.3 Vague Regulations for Obtaining STDB as an ISPO Certification Requirement and Replanting

As an ISPO pilot project, unfortunately, the Sari Makmur community did not receive a comprehensive explanation regarding administrative requirements and the process to obtain palm oil sustainable certification during the ISPO socialization process. The administrator of Mulia merely went about collecting documents administered by the company without understanding why. To be sure, ISPO implementation entails more than matching documents necessary for certification. At play here were local dynamics and different interpretations of obtaining documents such as STDB and SPPL. Leaders of the cooperative have admitted that obtaining these documents would cost around Rp1.3 million. In other words, a cooperative/organization will have to pay Rp650 million to accomplish the STDB and SPPL documents. Adding this to the cost of certification, in total, a cooperative would have to have around Rp700 million in its account.

This is a heavy cost to bear by the cooperative. The community expects the local government and the company to shoulder the cost. The community also maintains good relationships with the current district leader and member of local parliament to communicate this problem. This example happened in a neighboring village, which had also been targeted by PT. SLS for ISPO certification. The presence of local parliament representatives in Sari Makmur and its neighboring village could push for easier communication with leaders. The Sari Makmur community hopes that by communicating their problem, it will help the community to shift the costs of the STDB and SPPL applications to the company or local government. Moreover, the Mulia administration understands that in other provinces, such as Jambi, communities do not have to pay for these documents. In addition, the Sari Makmur community is a strong political supporter of the current district leader. Leaders of Mulia understand their political position and are confident in capitalizing on it during formal meetings with government representatives to raise their concerns. These community efforts show the leaders' abilities to use social capital to bargain with the government.

At the local government level, regulation for an STDB license mentioned in the Plantation Law and the regulation of the Ministry of Agriculture generated confusion in its practical implementation. In the context of an authorized agency, the plantation department in the local district is responsible for handling document collection and registration. This department did not impose costs for the document. This authority was transferred to the Board of Investment and Integrated Licensing Service (BPMP2T, *Badan Penanaman Modal dan Pelayanan Perizinan Terpadu*); however, there would be a cost charged for the document handling. At some point, the local government recognized the potential of STDB registration as a source of additional revenue, rather than as a vehicle for information dissemination that would help smallholder farmers carry out accurate spatial planning. The cost attached to STDB could also be found in other districts, such as Siak.

The STDB document is categorized by BPMP2T as a non-licensing document. Therefore, it can be inferred that the local government does not impose a cost on STDB arrangements. In practice, there is an operational cost that has to be borne by the plantation department as the responsible agency for issuing STDB documents before they are legally transferred to BPMP2T. An STDB application requires knowledge of the location coordinates as well as an on-site inspection by a team of experts from the plantation department. Unfortunately, the plantation department cannot cover this operational cost because there is no official local government allocation for this duty. Therefore, the operational problem should be communicated and coordinated with the plantation department, team of experts, farmers, and the company.

For an implementing agency such as the local plantation department, consolidating costs with the smallholder is the best solution. If they have to follow the application of ISPO rules in which the certification cost, especially for the smallholder, is obtained from the local government, it will take longer to process the ISPO certification. This is because allocating local government expenditure requires political consolidation. From the local government's perspective, consolidating the cost to obtain STDB and SPPL could open an opportunity for company involvement.

11.6   Conclusion

The paper has shown how legal requirements and processes to be fulfilled by both the smallholder and company have slowed ISPO implementation. The reluctance of the company and smallholder to participate in the tedious process does not necessarily imply any past legal wrongdoing on their part. In the case of the palm oil smallholders in Sari Makmur, the long process for ISPO certification is from a re-delineation of resource boundaries between the company and the community.

Even though both parties have been cooperating through the nucleus-plasma mechanism, the land swap in the past has resurfaced as a point of contention, with some groups in the community intending to break from the commitment by seizing an opportunity opened up by ISPO implementation. This problem has complicated the process of gaining STDB for the community as well as returning company land concession.

The STDB requirement for ISPO certification has been the intersection for the company and the community at which they made arrangements to pursue their mutual interests. For the company, community STDB ownership along with their land ownership will clear resource boundaries between company concession and community areas for palm oil cultivation. Therefore, it was not surprising when the company offered assistance to the community to collect the documents required in the application for ISPO certification. Meanwhile, the community expected that STDB ownership would open further assistance for the replanting process.

These show that the relationship between the company and the community is relatively fluid. The power imbalances in terms of capital and commodity prices do not shut down communication lines between them, especially in pursuit of win–win solutions. The company sees the Sari Makmur community as an independent and yet beneficial partner for their palm oil production cycle. Conversely, the community sees the company as a beneficial patron that can assure stable and fair prices for their palm oil output.

The cost of obtaining STDB is also an important matter for both the company and community. The community insists that the company bears the cost for STDB and the ISPO certification process, but the company thinks otherwise. Both parties occasionally discuss the matter through monthly meetings or at company events. Nevertheless, the community is making representations with the local government through linkages and networks with the district leader and the administration. The issue of cost for STDB results from the uncertainty in the ISPO implementation as to who ultimately bears the cost of the certification process.

Palm oil smallholders, especially those that are independent, are still facing problems with land legality. This occurs during the initial land clearing for plantation due to the overlapping of smallholder land ownership with a company's land use right or land clearing for a plantation on land forest status. This problem has limited the ISPO certification process for smallholders, which encounter difficulties in registering their smallholding or plantation, let alone gaining ISPO certification.

References

Bacon C (2005) Confronting the coffee crisis: can fair trade, organic, and specialty coffees reduce small-scale farmer vulnerability in Northern Nicaragua? World Dev 33(3):497–511. https://doi.org/10.1016/j.worlddev.2004.10.002

Bartley T, Smith SN (2010) Communities of practice as cause and consequence of transnational governance: the evolution of social and environmental certification. In: Djelic ML, Quack S (eds) Transnational communities: shaping global economic governance. Cambridge University Press, Cambridge, pp 347–374. https://doi.org/10.1017/CBO9780511778100.016

Beuchelt TD, Zeller M (2012) The role of cooperative business models for the success of smallholder coffee certification in Nicaragua: a comparison of conventional, organic and organic fair-trade certified cooperatives. Renew Agric Food Syst 28(3):195–211. https://doi.org/10.1017/S1742170512000087

BPS (Badan Pusat Statistik) (2016) Statistik kelapa sawit Indonesia 2016. BPS, Indonesia

Brandi C, Cabani T, Hosang C et al (2013) Sustainability certification in the Indonesian palm oil sector: benefits and challenges for smallholders. German Development Institute, Bonn. https://www.die-gdi.de/uploads/media/Studies_74.pdf. Accessed 7 Sep 2018

Bray DB, Sánchez JLP, Murphy EC (2002) Social dimensions of organic coffee production in Mexico: lessons for eco-labeling initiatives. Soc Nat Resour 15(5):429–446. https://doi.org/10.1080/08941920252866783

Cameron B (2017) Forest-friendly palm production: certifying small-scale farmers in Indonesia, 2011–2016. Princeton, Innovations for Successful Societies

Carlisle K, Gruby RL (2019) Polycentric systems of governance: a theoretical model for the commons. Policy Stud J 47(4):927–952. https://doi.org/10.1111/psj.12212

Cashore B, Auld G, Newsom D (2008) Governing through markets: forest certification and the emergence of non-state authority. Yale University Press, New Haven. https://doi.org/10.12987/9780300133110

Cramb R, McCarthy JF (eds) (2016) The oil palm complex: smallholders, agribusiness and the state in Indonesia and Malaysia. NUS Press, Singapore

Dove MR (2012) The banana tree at the gate: a history of marginal peoples and global markets in Borneo. NUS Press, Singapore

Ebeling J, Yasué M (2008) The effectiveness of market-based conservation in the tropics: forest certification in Ecuador and Bolivia. J Environ Manag 90(2):1145–1153. https://doi.org/10.1016/j.jenvman.2008.05.003

Elbers W, van Rijsbergen B, Bagamba F et al (2015) The impact of utz certification on smallholder farmers in Uganda. In: Ruben R, Hoebink P (eds) Coffee certification in East Africa: impact on farmers, families and cooperatives. Wageningen Academic Publishers, Wageningen, pp 53–82. https://doi.org/10.3920/978-90-8686-805-6_2

Erman E (2017) Di Balik keberlanjutan sawit: aktor, aliansi dalam ekonomi politik serifikasi Uni Eropa. Masy Indones 43(1):1–13

Frey BS, Obelhozer-Gee F (1997) The cost of price incentives: an empirical analysis of motivation crowding-out. Am Econ Rev 87(4):746–755

Gamson WA (2005) Movement impact on cultural change. http://havenscenter.org/files/MovementImpactOnCulturalChangeGAMSON1.pdf. Accessed 15 Jan 2018

Garrick DE, Villamayor-Tomás S (2017) Competition in polycentric governance system. In: Thiel A, Blomquist WA, Garrick DE (eds) Governing complexity: analyzing and applying polycentricity. Cambridge studies in economics, choice, and society. Cambridge University Press, Cambridge, pp 152–172. https://doi.org/10.1017/9781108325721.008

Guthman J (2004) Back to the land: the paradox of organic food standards. Environ Plan A 36(3):511–528. https://doi.org/10.1068/a36104

Hekkila T (2019) Conflict and conflict resolution in polycentric governance systems. In: Thiel A, Blomquist WA, Garrick DE (eds) Governing complexity: analyzing and applying polycentricity. Cambridge studies in economics, choice, and society. Cambridge University Press, Cambridge, pp 133–151. https://doi.org/10.1017/9781108325721.007

Hidayat NK, Glasbergen P, Offermans A (2015) Sustainability certification and palm oil smallholders' livelihood: a comparison between scheme smallholders and independent smallholders in Indonesia. Int Food Agribus Manag Rev 18(3):25–48

Hidayat NK, Offermans A, Glasbergen P (2018) Sustainable palm oil as a public responsibility? On the governance capacity of Indonesian Standard for Sustainable Palm Oil (ISPO). Agric Hum Values 35(1):223–242. https://doi.org/10.1007/s10460-017-9816-6

Hutabarat S (2017) Tantangan keberlanjutan pekebun kelapa sawit rakyat di Kabupaten Pelalawan, Riau dalam perubahan perdagangan global. Masy Indones 43(1):47–64

Jiwan N (2012) The political ecology of the Indonesian palm oil industry: a critical analysis. In: Pye O, Bhattacharya J (eds) The palm oil controversy in Southeast Asia. ISEAS Publishing, Singapore, pp 48–75. https://doi.org/10.1355/9789814311458-007

Komisi ISPO (2017) Daftar perusahaan penerima ISPO. Komisi ISPO, Jakarta

Koontz TM (2019) Cooperation in polycentric governance systems. In: Thiel A, Blomquist WA, Garrick DE (eds) Governing complexity: analyzing and applying polycentricity. Cambridge studies in economics, choice, and society. Cambridge University Press, Cambridge, pp 115–132. https://doi.org/10.1017/9781108325721.006

Li TM (2007) The will to improve: governmentality, development, and the practice of politics. Duke University Press, Durham

Marshall GR (2015) Polycentricity and adaptive governance. Paper presented at 15th Biennial Global Conference of the International Association for the study of the commons, Edmonton, 25–29 May 2015

McCarthy JF (2011) Processes of inclusion and adverse incorporation: oil palm and agrarian change in Sumatra, Indonesia. In: Borras SM, McMichael P, Scoones I (eds) The politics of biofuels, land and agrarian change. Routledge, Oxon, pp 247–276

McCarthy JF, Gillespie P, Zen Z (2012) Swimming upstream: local Indonesian production networks in "globalized" palm oil production. World Dev 40(3):555–569. https://doi.org/10.1016/j.worlddev.2011.07.012

Menteri Pertanian Republik Indonesia (2015) NOMOR 11/Permentan/OT.140/3/2015 tentang sistem sertifikasi kelapa sawit berkelanjutan Indonesia (Indonesian Sustainable Palm Oil certification system/ISPO). Kementrian Pertanian Republik Indonesia

Morrison TH (2017) Evolving polycentric governance of the great barrier reef. PNAS 114(15):E3013–E3021. https://doi.org/10.1073/pnas.1620830114

Ostrom E (1990) Governing the commons: the evolution of institutions for collective action. Cambridge University Press, Cambridge. https://doi.org/10.1017/CBO9780511807763

Ostrom V, Tiebout CM, Warren R (1961) The organization of government in metropolitan areas: a theoretical inquiry. Am Polit Sci Rev 55(4):831–842. https://doi.org/10.2307/1952530

Peluso NL (2016) The plantation and the mine: agrarian transformation and the remaking of land and smallholders in Indonesia. In: McCarthy JF, Robinson K (eds) Land and development in Indonesia: searching for the people's sovereignty. ISEAS-Yusof Ishak Institute, Singapore, pp 35–62

Pérez-Ramírez M, Ponce-Díaz G, Lluch-Coto S (2012) The role of MSC certification in the empowerment of fishing cooperatives in Mexico: the case of red rock lobster co-managed fishery. Ocean Coast Manag 63:24–29. https://doi.org/10.1016/j.ocecoaman.2012.03.009

Potter L (2016) Alternative pathways for smallholder oil palm in Indonesia: international comparisons. In: Cramb R, McCarthy JF (eds) The oil palm complex: smallholders, agribusiness and the state in Indonesia and Malaysia. NUS Press, Singapore, pp 155–188. https://doi.org/10.1355/9789814762106-007

Presiden Republik Indonesia (2020) NOMOR 44 TAHUN 2020 tentang sistem sertifikasi perkebunan kelapa sawit berkelanjutan Indonesia. Presiden Republik Indonesia

Rietberg P (2016) Barriers to smallholder RSPO certification: a science-for-policy paper by the SEnSOR programme. Wageningen University, Wageningen

Ruben R, Zuniga G (2011) How standards compete: comparative impact of coffee certification schemes in Northern Nicaragua. Supply Chain Manag 16(2):98–109. https://doi.org/10.1108/13598541111115356

Schröder NJS (2018) The lens of polycentricity: identifying polycentric governance systems illustrated through examples from the field of water governance. Environ Policy Gov 28(4):236–251. https://doi.org/10.1002/eet.1812

Silver H (1994) Social exclusion and social solidarity: three paradigms. Int Labour Rev 133:531–578

Somerville P (2000) Social relations and social exclusion: rethinking political economy. Routledge, London. https://doi.org/10.4324/9780203184974

Stephan M, Marshall G, McGinnis M (2019) An introduction to polycentricity and governance. In: Thiel A, Blomquist WA, Garrick DE (eds) Governing complexity: analyzing and applying polycentricity. Cambridge studies in economics, choice, and society. Cambridge University Press, Cambridge, pp 21–44. https://doi.org/10.1017/9781108325721.002

Suharto R, Husein K, Sartono et al (2015) Studi bersama persamaan dan perbedaan sistem sertifikasi ISPO dan RSPO. ISPO/RSPO, Jakarta/Kuala Lumpur

Suradisastra K (2006) Agricultural cooperative in Indonesia. Paper presented at 2006 FFTC-NACF international seminar on agricultural cooperatives in Asia: Innovations and opportunities in the 21st Century, Seoul, 11–15 September 2006

Thiel A, Blomquist WA, Garrick DE (eds) (2019) Governing complexity: analyzing and applying polycentricity. Cambridge studies in economics, choice, and society. Cambridge University Press, Cambridge. https://doi.org/10.1017/9781108325721

Thomas M, Buckmaster L (2010) Paternalism in social policy: when is it justifiable? Research paper 8. Parliament of Australia, Canberra

van der Enden S (2013) Smallholders and sustainable palm oil production: a better understanding between policy arrangements and real life practices: a case study of the Siak smallholders site, Riau Province, Sumatra. Dissertation,. Wageningen University

Vandergeest P (2007) Certification and communities: alternatives for regulating the environmental and social impacts of shrimp farming. World Dev 35(7):1152–1171. https://doi.org/10.1016/j.worlddev.2006.12.002

Varkkey H (2016) The haze problem in Southeast Asia: palm oil and patronage. Routledge, Oxon and New York

Werle R (2001) Standards in the international telecommunications regime. HWWA Discussion Paper no 157, Hamburg Institute of International Economics, Hamburg

Widyatmoko B (2019) The implementation of Indonesian Sustainable Palm Oil certification (ISPO): opportunity for inclusion of palm oil smallholder in Riau Province. Dissertation,. Kyoto University

Wolf ER (2006) Incorporation and identity in the making of the modern world. In: Kirsch M (ed) Inclusion and exclusion in the global arena. Routledge, New York and Oxon, pp 209–224

Printed in the United States
by Baker & Taylor Publisher Services